# Decision Engineering

*Series Editor*

Dr Rajkumar Roy
Department of Enterprise Integration
School of Industrial and Manufacturing Science
Cranfield University
Cranfield
Bedford
MK43 0AL
UK

*Other titles published in this series*

*Cost Engineering in Practice*
John McIlwraith

*IPA – Concepts and Applications in Engineering*
Jerzy Pokojski

*Strategic Decision Making*
Navneet Bhushan and Kanwal Rai

*Product Lifecycle Management*
John Stark

*From Product Description to Cost: A Practical Approach*
*Volume 2: Building a Specific Model*
Pierre Foussier

*Decision-Making in Engineering Design*
Yotaro Hatamura

*Intelligent Decision-making Support Systems: Foundations, Applications and Challenges*
Jatinder N.D. Gupta, Guisseppi A. Forgionne and Manuel Mora
Publication due April 2006

*Metaheuristics: A Comprehensive Guide to the Design and Implementation of Effective Optimisation Strategies*
Christian Prins, Marc Sevaux and Kenneth Sörensen
Publication due December 2006

*Context-aware Emotion-based Multi-agent Systems*
Rajiv Khosla, Nadia Bianchi-Berthouse, Mel Seigel and Toyoaki Nishida
Publication due July 2006

Pierre Foussier

# From Product Description to Cost: A Practical Approach

## Volume 1: The Parametric Approach

With 83 Figures

Springer

Pierre Foussier, MBA
3f, 15, rue des Tilleuls
78960 Voisins le Bretonneux
France

British Library Cataloguing in Publication Data
Foussier, Pierre
   From product description to cost: a practical approach
   Volume 1: The parametric approach. - (Decision engineering)
   1. Production planning - Mathematical models 2. Start-up costs
   Mathematical models 3.New products - Decision-making
   I. Title
   658.1'552'015118
   ISBN-10: 185233973X

Library of Congress Control Number: 2005937146

Decision Engineering Series ISSN 1619-5736
ISBN-10: 1-85233-973-X     Printed on acid-free paper
ISBN-13: 1-85233-973-9

© Springer-Verlag London Limited 2006

Apart from any fair dealing for the purposes of research or private study, or criticism or review, as permitted under the Copyright, Designs and Patents Act 1988, this publication may only be reproduced, stored or transmitted, in any form or by any means, with the prior permission in writing of the publishers, or in the case of reprographic reproduction in accordance with the terms of licences issued by the Copyright Licensing Agency. Enquiries concerning reproduction outside those terms should be sent to the publishers.

The use of registered names, trademarks, etc. in this publication does not imply, even in the absence of a specific statement, that such names are exempt from the relevant laws and regulations and therefore free for general use.

The publisher makes no representation, express or implied, with regard to the accuracy of the information contained in this book and cannot accept any legal responsibility or liability for any errors or omissions that may be made.

Typesetting: Gray Publishing, Tunbridge Wells, UK
Production: LE-TEX Jelonek, Schmidt & Völcker GbR, Leipzig, Germany

Printed in Germany

9 8 7 6 5 4 3 2 1

Springer Science+Business Media
springer.com

*To my daughter Pierrine,
without whom this book could not have been written.*

# Preface

Traditionally industry estimates costs when the drawings are ready. From the drawings the operations sheets are prepared (these sheets list all the activities, or operations, which will have to be carried out in order to manufacture the equipment); then, for each operation, one can decide that such a machine, because it works at a specific speed, will have to be used during a given time (which will become the "allocated time" for the operation). Knowing the cost of the machine, its depreciation, the costs of the energy, of the fluids, of the operator, etc. one can establish the cost of the operation. The cost of the product is then estimated as the sum of the costs of the operations, plus some other costs (tooling, management, quality control, etc.).

It is not possible nowadays to go on with this traditional approach for several reasons:

1. The cost is estimated far too late; nowadays decision about the production must be made very early, long before drawings are available. One cannot anymore prepare the drawings without having first studied the profitability of the product, which implies to have estimated its costs long before drawings are produced.
2. The process is time consuming and far too costly. It should be considered, because it is necessary, only if the profitably is well established and when alternatives, trade-offs analysis, etc. have been made.
3. If several alternatives have too be considered in the early stages of a project (and there are always some), the choice has to be made long before the detailed studied are made. And this choice depends – at least partly – on the cost estimates of the various alternatives.
4. Modern cost management uses disciplines such as the design to cost (DTC) or the cost as an independent variable (CAIV), which are all based on permanent re-estimates.
5. It is important at the early stages of a project to be able to carry out trade-offs analysis. The purpose of the analysis is to adjust, modify, delete, etc. the functional specifications in order to get the best compromise between competing technical performances and cost for the customer. These analysis obviously require early – and sometimes very early – cost estimates, most often at the functional level.

The solution is, of course, to be able to estimate directly from the product description (or preferably even from product specifications), without having to prepare the operations sheets. The description has sometimes to be only a preliminary description.

How can we do that? This is the purpose of this book. The idea is to look for the "cost drivers" of the product (functional and/or physical) and to establish relationships between these cost drivers and the cost. These relationships are called "cost models". A cost model is a mathematical (but it can also be some graphs or simple tables) representation of how the cost changes when each cost driver changes.

Do we need a "complete" description of the product to be able to estimate its costs? The answer of this question depends on the kind of model you are looking for:

- The simplest models just compare between themselves products which can be compared. Comparable products are products which fulfill the same function(s) at different levels, using different materials, etc. These products will be said, "belonging to the same product family". Inside a product family you do not need to fully describe the products: you just have to describe the few points by which they differ (the size, for instance, being always an important point). The models which will be built for these "product families" are called "specific models". They are "specific" because they can only be used for the product families they were built for.
- The most complex models can estimate any product without comparing it – at least explicitly, because we will see it is not always the case in practice – to any other product. Quite obviously these models are far more difficult to prepare. These models are called "general models". They are "general" because they can be used for estimating the cost of any product, within certain limits of course.

This book is mainly dedicated to the specific models, for several reasons:

1. Any company has historical data (by data we mean the set "cost + product description": a cost without the product description has no value at all) and prefer – this is quite understandable – to "extrapolate" the cost of a new product from its own data, instead of using a general model on which it has little control. The credibility (this is an important word in cost estimating) will certainly be much higher in the first case than in the second.
2. The method which can be used for establishing such a specific model is now well known and properly formalized. It can be used by anybody and will give good results – or at least controllable results – if it correctly applied.

Nevertheless some comments, which should be looked at as an introduction to the subject, are given about the general models, in order for the reader to understand the various ways which can be used for building such models.

This book aims at being practical. "Practical" does not mean lack of formulae. In the author's mind, it means discovering concepts from practical situations. Therefore most of the concepts are present in this book, but they are always explained and documented on the basis of numerical exercises: it is well known that theories are easier to understand if they are illustrated with examples.

The origin of this book comes from two observations: the misunderstanding and often the rejection, by many potential users who would take benefits from it, of parametric (this is the adjective often given to cost modelization) cost estimating on one hand, some lack of rigor of several users of this technique on the other hand (these users are convinced that this technique can help them, but they do have the time to study it, as its description is scattered in many books). It is very likely that the second point fosters the first one: lack of rigor discourages many observers of using such a technique. We have to break this link: properly used parametric cost estimating is a powerful and very cost effective technique, but, as for every technique, it has limitations.

Rigor is the key word in cost estimating which has an important need for credibility. Credibility can only be found in an agreed upon way of using a technique. This means that we all have in the cost estimating community to agree on a corpus of knowledge, and to recognize its limitations. This book is a contribution towards this end.

"Parametrics" as such is a very old technique that all the technicians and engineers learnt when they were at school and we may add that it is nowadays the normal way when dealing with engineering forecasts. When you write, for instance $pV = RT$ (Boyle's law for perfect gases), you are thinking "parametrically": you link the pressure $p$, the volume $V$ and the temperature $T$ (which are all called "parameters") by a relationship using a constant $R$, the "perfect gas constant". The relationship may be called a law, or a formula, or an equation, or an algorithm, or a model, …; the word is not important, but the concept is. In this book we will use extensively the word "model" because it clearly describes what we are trying to do (modelization, by some mathematical relationships, of something we observe in nature). This Boyle's law is perfectly usable, and very useful, in most situations, but it has its own limitations (when the pressure becomes too high it has to be replaced by the Van der Waals equation): it is nevertheless an acceptable and fast way to predict one of the variables in most situations.

In cost estimating, we work exactly the same way: we try to establish relationships between the cost and some physical variables or parameters (we are not sure why this word replaced, in the cost domain, the name of "variable" which is more explicit and understandable; maybe it was chosen[1] because it looks more scientific? Nevertheless it is there and we will have to use it): all cost estimators, whatever they estimate, whatever the way they estimate, apply the same principle. It can be applied very simply (when you say, for instance, that the cost for making a trench is equal to the unit cost – what we call the "specific cost" – multiplied by the trench size) or in a more complex way (taking into account, probably in a non-linear way, the type of soil, the type of machinery, etc.), but there is no other way to make a cost estimate, apart a direct comparison with some past experience. Therefore parametrics is a very, very common technique: the word was probably a bad choice, because it frightens many potential users, but what it covers is very basic.

Why is then the technique so often rejected by people who use it day after day outside the cost environment? We saw at least six basic reasons:

1. The fact that there is no consensus in the world of "parametricians" about what should be the right way(s) to use it. Parametric cost estimating is – a little bit – taught, statistical analysis is largely taught, but it is extremely rare to see a lecture on how statistics can be used for cost estimating, about the limits of these techniques, on how to judge the quality of a model, on how to audit it, etc. Some theoretical improvements are sometimes done by very clever people, but they generally forget the basis on what is built the statistical analysis of costs they make.
2. The existence of two completely different parametric techniques: one is based on a direct analysis of past experiences (known costs), the other one on "conceptual" modelization – this is most often (not always) the case for commercial

---

[1] From what I remember from my studies in mathematics, a parameter was an auxiliary variable which was used to introduce a (small) change in the main relationships … This is not the case here: a parameter has the same status as a variable.

models – for which the relationships with past experiences, although it does exist, is not so clear. This book describes both techniques under different parts (Volume 2 deals with the first one; Part III of this Volume 1 with the second one). The distinction is very clear, but many people, including the sellers of these commercial models, present them as if they were identical: they are not, and we will see they cannot be; but the confusion probably frightens many potential users.

3. A lack of rigor of some users. For the first technique, one could expect the technique to be stabilized, even if some improvements are always possible, but it is not: we saw (even in international conferences) presentations that would be immediately rejected in any other engineering activity. It is therefore not a surprise if the technique has a poor image. What about the other technique? In this second technique, the role of the user is extremely important: the same conceptual model can be applied in different ways, giving different results … When we hear someone say "the model says that the cost is …", we immediately want to convert it into something such as "using the model this way, with these inputs, we found such a result!".
4. The belief that everybody, any company, is specific in the domain of cost (this belief does not exist, fortunately enough, in other engineering sciences). The consequence is obvious: as we are all specific, we have nothing to learn from other companies. These people are partly right: they are specific in the type of products – or software, or anything else – they make; but they are not specific in the way they do it, except when a new way is found (but this new way does not remain "new" for a long time). A true conceptual model is precisely trying to modelize the way people work, not on what they work. Therefore many people could take benefits of it.
5. The belief, which is extremely strong, that cost estimating requires a complete knowledge on how a company works, on the machine tools it uses, on how it organized, etc.; in other words, cost estimating can only be done in a very detailed way. And parametrics is often presented as a shortcut! We do not challenge the first sentence: cost estimating can be done in a detailed way and it has to be done if the manufacturing process completely changes from one product to another one; we only challenge the word "only": cost estimating can be done, in most practical situations, in another way, as it will be described in this book and this other way can be extremely useful at a given stage – not always! – of a project.
6. Parametrics is sometimes presented as an extension of a "cost per kilogram" technique! It is not: we totally agree on the fact that "cost per kilogram" does not work, unless we compare products so close to each other that no parametric tool is necessary. The reason why it does not work will be given in this book.

This book has three major objectives:

1. Present the different ways of "parametrically" forecast costs,[2] with their advantages, their inconveniences, their limitations.
2. Give nearly all mathematical procedures which are useful for parametrically forecasting costs.

---

[2] Instead of cost, one could speak about men-hours, or number of persons for doing a specific task, such as project management, etc.

3. Introduce the logic to audit the way the techniques are used, as a process first (which means generally speaking inside a company), secondly as a tool to generate estimates.

Many examples are given in order to make the reader able to use the concepts, discuss them and apply them to real situations.

Many formulae are given. As the purpose of the book is not to study the theory, but to practically present the ideas in order that the reader may use them immediately, demonstrations of the formulae are not generally given. Demonstrations can easily be found in many theoretical books of which titles are given.

The book is divided into two volumes.
Volume 1 includes four parts:

- Part I is an **introduction** to cost estimating:
  - In terms of definition of several concepts and of cost measurement (the book is not a lecture on cost accounting, but any cost analyst should have some understanding about how costs are "measured" and the limitations of the published costs).
  - In terms of presenting the two major techniques of "parametric" cost estimating: this is the purpose of Chapter 3, which is therefore an important chapter of the book. Both techniques build "models", but these models are very different; they are called "general models" (they will be described in Part III) on one hand, "specific models" (they will be described in Volume 2) on the other hand.
- Part II deals with the important subject of preparing the data before any utilization – either for general or specific models – can be made of them. Preparation of data is important because the only way we, as human beings, are able to forecast the future is by extrapolating the results of past experiences. This implies that these results have to be compared and comparisons can only be done on comparable data. Making the data comparable is sometimes called their **normalization**. This is a good word. It implies that, once a relationship has been found between the normalized data and that the cost of new product has been computed, this result has to be denormalized.
- Part III describes the basic concepts of "**general**" cost estimating **models**. These models are not purely built on statistical analysis of existing data; data are obviously used to build them, but preconceived ideas are as important as the data themselves. There is no documented procedure for building such models and this is what makes their creation difficult and the comparison between these models sometimes arduous. For the cost estimator, using such a model starts with a "translation": his/her company and the projects managers have their vocabulary and the model (built by persons who do not know how the model will be used and for what purpose) has its own. Both do not coincide, which makes a translation necessary. The more "abstract" the model, the more sources of errors the translation can generate: using a general model needs a good understanding of the model logic.

  The major advantage of general models compared to specific models is that they theoretically need very little data from the company. The price to pay for that is that you must get some definition about the product you want to estimate.
- Part IV discusses the use of cost models (either general or specific) and introduces **risk analysis** as it can be used during parametric cost estimating.

It describes the probabilistic and the possibilistic approaches; it will be shown that the first one is generally the less adapted technique and how it can lie to the decision-maker. It also comments on how to audit parametric cost estimating.

Volume 2 deals with the building of "**specific**" cost estimating **models** (sometimes also called "in-house models"). Not so many persons are able to build general models, but any company – if it is properly organized for this goal – can build specific ones. Therefore the reader is strongly encouraged to start parametric cost estimating following this path. The techniques for building such models are now well understood (even if they can still be improved) and the cost analyst has just to follow the documented procedures. During his/her work, he/she will have to make decisions, which require understanding the procedures.

Understanding the procedures is the key word for creating successful specific models. For this reason all these procedures are fully described and this description requires a full volume. Classical methods and new ones (such as the "bootstrap") will be described and illustrated.

There is no translation effort for building and using specific models: these models use data generated by the company (and therefore data collection is a major element of the process), they also use company vocabulary. For this reason the credibility, a key word in parametric cost estimating, is easier to achieve than with general models.

The major constraint for building specific models is to get as many data as possible from past projects: the more data you have, the more confident you will be in the cost forecasts made by the model.

This Volume 2 includes five parts:

- Part I briefly introduces several concepts which are used throughout the volume and presents the main logic on which is based the construction of specific models.
- Part II details all the analysis which must be done on the data – supposed having been normalized as described in Volume 1 – before any model building can start. The purpose of this data analysis is there to make sure that no problem related to the data themselves can endanger the creation of the specific model which is looked for.
- Part III is a prerequisite for model building. A model cannot – and should not – represent the data with an infinite precision: it has to eliminate errors, mistakes, misinterpretations, fluctuations, etc., which are inherent to human activities but must not be included in a model in order for such a model to be easily used at the time it has to be. This is done via the choice of a "metric". Once this is done, what we call the "dynamic center of the sample cost data", which is the heart of the specific model, can be established.
- Part IV investigates what is left over – what we call the "residuals" – when the process described in the preceding part is finished. The study of these residuals is as important as the search for the dynamic center.
- Part V is really the construction of the model: it describes how what has been found from the sample data can be applied to the whole "population" from which the sample was "drawn".

Cost estimating requires 30% of data, 30% of tools and 40% of judgment and thinking, with a minimum of 80% in total. EstimLab™ – with which most of the computations which illustrate this book have been performed – was designed to get all these 30% of tools with a minimum of effort, freeing time for collecting data and making use of judgment, which is always the most important component in cost estimating.

I was trained first as a mathematician, then as a scientist and eventually as a business manager. From my first training, I kept a taste for concepts, demonstrations, checking the hypotheses before I apply a theorem, etc. From the second one, I kept a great respect to the facts: the first question we have to ask when dealing with a new problem is: What are the facts? From the third one, I realized that concepts and facts have to produce a result useful for the company I work for.

I think cost estimates should be based on these three foundations.

No prerequisite is required to read this book. It was written to be self-sufficient.

This book is not a book on mathematics or even statistics. Several concepts it uses are generally presented in statistical manuals. But our purpose is not to do statistics on the few data we generally have when dealing with cost, but to analyze these data and to extract from them what can be used for making forecasts.

Mathematics and statistics are there just to help us, not as an objective by themselves.

I would like to thank Gilles Turré who wrote an important chapter of this book which will be a valuable interest to all the readers interested in mass production. His knowledge of the subject and his ability to present it contribute to the completeness of the methods presented to the cost analyst.

*Paris*  *Pierre Foussier*
*February 2005*

# Contents

Notations .......................................................... xxi
What you need to know about Matrices Algebra ....................... xxvii

## Part I  The Framework

### 1  Cost Estimating: Definition ........................... 3
1.1  Some Definitions ..................................... 3
    1.1.1  What Is Cost? .................................. 3
    1.1.2  About Price .................................... 6
1.2  What Is Cost Estimating? ............................. 7
1.3  The Fundamental Theorem About Forecasting ........... 10
1.4  A Few Comments ...................................... 11

### 2  Cost Measurement ................................... 13
2.1  The General Framework of Cost Measurement .......... 15
2.2  Two First Distinctions: Fix and Variable, Direct and Indirect  17
2.3  Direct Costing ....................................... 19
2.4  Full or Absorption Costing ........................... 20
    2.4.1  "Rough" Full Costing ........................... 21
    2.4.2  The "Standard" Solution ........................ 24
2.5  The Method ABC ...................................... 26
2.6  What Can the Cost Analyst Expect? .................... 28

### 3  Overview of Cost Estimating ......................... 31
3.1  Product Description and Approach .................... 32
3.2  The Usual Methods ................................... 34
    3.2.1  Heuristics ..................................... 38
    3.2.2  Expert Judgment ............................... 40
    3.2.3  Data Organization and CBR ..................... 40
3.3  Modelization (Parametric Cost Estimating) ............ 41
    3.3.1  Definitions .................................... 41
    3.3.2  A Brief History ................................ 42
    3.3.3  The Foundations ............................... 43
    3.3.4  Several Levels of Consistency ................... 44
    3.3.5  Two Different Types of Models .................. 45

### 4  "Elementary" Cost Estimating ....................... 47
4.1  Only One Data Point Is Available ..................... 47
    4.1.1  No Other Information Is Available .............. 47

|  |  |  |  |
|---|---|---|---|
|  |  | 4.1.2 Other Information Is Available ................... 49 |  |
|  | 4.2 | Two Data Points Are Available .......................... 53 |  |
|  |  | 4.2.1 No Other Information Is Available ................ 53 |  |
|  |  | 4.2.2 Other Information Is Available ................... 55 |  |
|  | 4.3 | Two Data Points Are Available and Two Parameters Must be Considered ........................................ 55 |  |

## 5 An Introduction to Parametrics for the Beginner ..... 59
5.1 The Data .............................................. 60
5.2 Analysis of the Data .................................. 62
5.3 What Are the Relationships Between the Price and the Technical Variables? .................................. 63
    5.3.1 Relationships Between Price and One-Functional Characteristic ................................. 63
    5.3.2 Relationship Between Price and Several Characteristics ................................. 68
    5.3.3 Relationships Between Price and Mass ........... 71
    5.3.4 Is There a Correlation Between Mass and Seats × Range? ................................. 72
5.4 What Is the Quality of the Formula? .................. 73
    5.4.1 Examining the Residuals ....................... 73
    5.4.2 Quantifying the Quality of the Formula ......... 74
5.5 Estimating the Price of a New Aircraft ................ 77

## Part II  Preparing the Data

## 6 Data Collection and Checking ...................... 83
6.1 Collecting "Raw" Data ................................ 84
    6.1.1 What Do We Mean by Data? ..................... 84
    6.1.2 The Point of View of the Cost Analyst ............ 85
    6.1.3 A Word of Caution ............................. 86
    6.1.4 What Database? ............................... 86
6.2 Data Clarification .................................... 87
    6.2.1 Products Definition ............................ 87
    6.2.2 About the Activities ........................... 88
    6.2.3 About the Over-Costs .......................... 89
    6.2.4 Cost Definition ............................... 89
    6.2.5 Expert Judgment .............................. 90
6.3 Conclusion ........................................... 93

## 7 Economics ......................................... 95
7.1 Normalization in the Same Currency ................... 95
    7.1.1 Changing the Economic Conditions of One Expense ................................... 96
    7.1.2 Changing the Economic Conditions of a Total Project Cost .................................. 97
7.2 Between Two Countries ...............................101
7.3 Conclusion ..........................................104

## Contents

**8  The Cost Improvement Curves** .......... **105**
- 8.1  Introduction .......... 105
  - 8.1.1  Getting the Facts? .......... 109
  - 8.1.2  Finding the General Trend(s) .......... 110
  - 8.1.3  Introducing Parameter(s) .......... 111
  - 8.1.4  Establishing the Algorithm .......... 112
- 8.2  The Continuous Production: The Simple or "Pure" Models .......... 113
  - 8.2.1  Wright's Law (Deals with Average Time) .......... 113
  - 8.2.2  Crawford's Law (Deals with True Unit Time) .......... 117
  - 8.2.3  Comparing Wright's and Crawford's Laws .......... 119
  - 8.2.4  In Practice .......... 121
- 8.3  The Continuous Production: More Complex Models .......... 123
  - 8.3.1  The Broken Line Model .......... 123
  - 8.3.2  Other Laws .......... 126
- 8.4  More Complex Situations to Modelize .......... 128
  - 8.4.1  Modifying the Production Methods .......... 128
  - 8.4.2  Interrupted Production .......... 129
- 8.5  Applications .......... 131
  - 8.5.1  Normalization .......... 131
  - 8.5.2  Denormalization .......... 132
- 8.6  Conclusion .......... 132

**9  Plant Capacity and Load** .......... **133**
- 9.1  A Few Definitions .......... 133
  - 9.1.1  What Is a Product? What Is Its Size? .......... 133
  - 9.1.2  Parameters of a Production .......... 134
  - 9.1.3  Plant Capacity: Dedicated and General Capacities .......... 135
  - 9.1.4  Practical Summary of the Capacity Parameters .......... 137
  - 9.1.5  About the Cost .......... 140
- 9.2  Cost of Production as a Function of the Size .......... 141
  - 9.2.1  The Power Law or the Chilton's Rule .......... 141
  - 9.2.2  The Power Law for Specific Costs .......... 144
- 9.3  Relationship Between Plant Size and Costs .......... 144
  - 9.3.1  Cost of a Production as a Function of the Plant Capacity .......... 144
  - 9.3.2  Cost of the Products as a Function of the Size of Their Manufacturing Unit .......... 146
  - 9.3.3  Example .......... 149
- 9.4  How Does the Load (of a Plant with a Stable Capacity) Influence the Production Cost? .......... 149
  - 9.4.1  What Is the Load? .......... 149
  - 9.4.2  The Causes of Variation of the Production Cost According to the Load .......... 150
  - 9.4.3  The Total Production Cost as a Function of the Quantity $Cst_{t/yr}(Qty_{yr})$ .......... 151
  - 9.4.4  Unitary Costs According to the Load of Plant .......... 157
- 9.5  In Practice .......... 164
  - 9.5.1  The Different Points of View .......... 164
  - 9.5.2  A Few Other Comments .......... 165

| | | |
|---|---|---|
| 10 | Other Normalizations | 167 |
| | 10.1 Normalization Related to the Plant Localization | 167 |
| | 10.2 Normalization Related to the Position in the Life Cycle of a Product | 167 |

| | | |
|---|---|---|
| Part III | **About General Models** | |
| 11 | Definition of a General Model: Example | 175 |
| | 11.1 Definition of a General Model | 175 |
| | 11.2 An Example of a General Model: Value Estimator | 179 |
| |     11.2.1 Product Description | 179 |
| |     11.2.2 Quality Level | 180 |
| |     11.2.3 Results | 182 |
| |     11.2.4 Conclusion | 184 |
| 12 | Building a General Model | 185 |
| | 12.1 The Three Types of General Models | 186 |
| | 12.2 The Price to Pay for Generality | 187 |
| |     12.2.1 About the Size | 187 |
| |     12.2.2 The Quality Level | 188 |
| |     12.2.3 The Time | 189 |
| | 12.3 Type 1 Models | 189 |
| | 12.4 Type 2 Models | 190 |
| |     12.4.1 Type 2 Models: Distinguishing Between "Laws" and "Initial Conditions" | 191 |
| |     12.4.2 A Few Difficulties When Using a Type 2 Model | 192 |
| |     12.4.3 Conclusion | 194 |
| | 12.5 Type 3 Models | 194 |
| | 12.6 Other Points | 195 |
| 13 | New Concepts in General Modelization for Hardware | 197 |
| | 13.1 Introduction | 198 |
| | 13.2 Describing the Product | 199 |
| |     13.2.1 The Product Size | 199 |
| |     13.2.2 Product Description Beyond the Size: The ITF | 203 |
| |     13.2.3 Quantification | 206 |
| | 13.3 The Quality Level | 210 |
| |     13.3.1 What Do We Mean by Quality Level? | 210 |
| |     13.3.2 Quantifying the Quality Level: A Basic Approach | 211 |
| |     13.3.3 A More Complete Approach | 213 |
| | 13.4 About the Materials | 214 |
| |     13.4.1 About the Machinability of Materials | 216 |
| | 13.5 The Manufacturing Process | 216 |
| | 13.6 The Basic Relationship Between Size and Manufacturing Cost | 216 |
| |     13.6.1 The Type of Formula | 217 |
| |     13.6.2 The Type of Production | 217 |

|  |  |  |
|---|---|---|
| | 13.7 | Adjusting the Model to the Manufacturer: The Concept of Industrial Culture .................. 218 |
| | | 13.7.1 How the Culture Can be used in a Model? .... 222 |
| | | 13.7.2 Conclusion .............................. 222 |
| | 13.8 | A Last Word ..................................... 223 |

# 14 Modelization in Other Classes ..................... 225
14.1 Software ......................................... 225
    14.1.1 About the Size ........................... 225
    14.1.2 Relationships Between Cost and Size ........ 227
    14.1.3 Taking into Account the Other Cost Drivers .. 228
14.2 Building ......................................... 229
    14.2.1 The Building Size ......................... 231
    14.2.2 Feasibility Factors ......................... 231
    14.2.3 The Quality Level ......................... 231
    14.2.4 Localization Factors ....................... 232
14.3 Tunnels .......................................... 232
    14.3.1 About the Size ........................... 233
    14.3.2 About the Rock .......................... 233
    14.3.3 The Excavating Technique ................. 233
    14.3.4 Several Costs ............................ 234

# 15 A Word About the Future ........................ 235
15.1 Measuring Entropy ............................... 235
15.2 Application to Manufacturing ..................... 236
15.3 About Materials .................................. 236
15.4 About Construction ............................... 236
15.5 About the Maintenance ........................... 236

# Part IV Using Models: Parametric Models Within a Business

# 16 Using a Specific Model ........................... 241
16.1 The Classical Approach ........................... 242
    16.1.1 Using one Parameter Only ................. 243
    16.1.2 Using Several Parameters .................. 247
    16.1.3 What Can be Done When the Number of Data Points Is Limited? .................... 247
16.2 The Modern Approach ............................ 248

# 17 Using a General Model ........................... 249
17.1 Using a Type 1 Model ............................ 250
17.2 Using a Type 2 Model ............................ 250
17.3 Using a Type 3 Model ............................ 251

# 18 Introduction to "Risk" Analysis ................... 253
18.1 Introduction ..................................... 254
    18.1.1 A Definition .............................. 254
    18.1.2 The Components of Risk ................... 256
18.2 Taking into Account the Imprecision of the Model Inputs ............................................ 259

|  |  | 18.2.1 | What Do We Mean by "Imprecision"? ........ 259 |
|  |  | 18.2.2 | Introduction to the "Fuzzy Numbers" ........ 260 |
|  |  | 18.2.3 | Using Fuzzy Numbers Inside Cost Models .... 266 |
|  | 18.3 | Using Discrete Probabilities ....................... 266 |
|  | 18.4 | Using the Continuous Probabilities ................ 267 |
|  |  | 18.4.1 | Using the Moments ...................... 268 |
|  |  | 18.4.2 | Using the Monte-Carlo Method ............ 278 |
|  | 18.5 | Mixing Both Methods ............................ 281 |

## 19 Auditing Parametrics ............................ 283
   19.1 Short-Term Auditing: About Using Models ......... 286
        19.1.1 Using Specific Models ..................... 286
        19.1.2 Using General Models ..................... 287
   19.2 Medium-Term Auditing: About Preparing Models .... 288
        19.2.1 About Specific Models ..................... 288
        19.2.2 Using General Models ..................... 289
   19.3 Long-Term Auditing: About the Process ............. 289

Bibliography ...................................................... 293
Index ............................................................. 295

# Notations

Cost estimation handles information.

This information is generally presented into tables of figures. In order to discuss about these tables, to analyze their structure, to establish relationships between them, etc., it is convenient to use symbols to represent them. The definition of these symbols is given in this section. We try to use symbols which are – by using simple rules – easy to remember. Most of them are common with the majority of the authors; a few of them, when experience showed the symbols generally used in the literature maybe confusing, are different.

Information is relative to objects or products (or "individuals" in statistical books). It is conveyed by variables.

The sample is the set of objects for which we know the value of the variables. The population is the set of objects, of which number is supposed to be infinite, for which we want to get a cost estimate and from which the sample is supposed to be "extracted".

## The "Individuals" or Products

The methods developed in this book can be applied to any set of objects. An element of a set is called an "individual" (the term reminds that statistical methods were developed for studying populations).

However, as its title mentions it, this book is principally dedicated to cost analysts and cost estimators. The subject of interest of these persons will be called "products".

The term "product" is therefore used here to designate anything we are interested in. It can be a piece of hardware, a system, a software, a tunnel, a bridge, a building, etc. Generally speaking, it is something which has to be designed and/or produced and/or maintained and/or decommissioned.

Products generally have names. In order to remain as general as possible, one will assign a letter to each one: they will be designated by capital letters such as $A_1$, $A_2, A_3, \ldots, A_i, \ldots, A_I$ ($A$ for "article") the index[3] "$i$" being reserved for the products.

The number of products we may have to deal with simultaneously (the meaning of this adverb will become clear in the following chapters) will be called $I$.

A product is characterized by a set of variables.

## The Variables

A **variable** is something which can be quantified (in such a case it is called a "*quantitative*" variable, the quantity being a number such as 27, or 34.5 or even – 12.78) or

---

[3] When indexes are used, it is very convenient – and easy to remember – to use a small letter for the current index and the same letter – this time in capital – for the upper limit of this index.

on which one can affect an attribute (in such a case it is called a "*qualitative*" variable; the quality being an adjective such as superior, good, poor or even a sentence, such as made by machine C, machine D, ... , or even sometimes an integer);[4] the attribute can be "*objective*" (if it expresses a fact, such as the material used, or the manufacturer) or "*subjective*" if it expresses an opinion (such as little complex, very complex ... the adjective "complex", or sometimes "difficult", being rather frequent for expressing an opinion about the nature of a product).

A **quantitative variable** must have a unit (such as kilogram, meter, inch, euro, dollar, or simply "quantity", generally simplified in "qty"). A limit can be imposed on the set of values it may take. It is supposed to be continuous, even it is not really, such as qty.

A **qualitative variable** refers to a limited set of attributes or modalities such as (manufactured by A or B) or (blue, green, yellow) or (very low, medium, high, very high).

A variable will always be symbolized by a capital $V$. This capital $V$ will only be used to represent the variable as such (e.g. one can say $V$ represents the product mass; sometimes one can even say – it is not really correct but generally accepted – $V$ "is" the mass), not its value.

Several variables will have to be used. Each variable has a name, such as the mass, or the power, or the speed, ...; for practical reasons they will nevertheless be represented by the capital letter $V$ with an index, such as $V_0, \ldots, V_j, \ldots, V_J$ the index "$j$" being reserved for the variables. In this book, dedicated to forecasting a value (in principle the cost, but the methods can be applied to practically anything), *two types of variables* must be distinguished:[5]

1. The "explicative" or "causal" variables, which are the variables that are known when a forecast has to be made; these variables are generally called "parameters". They will be represented by an index equal or superior to 1. Example: $V_2$. The number of these variables will be called $J$. $V_0$ will be used for a constant, as cost may have a constant part, which has of course also to be determined.
2. The "dependent" or "explained" variable, which is the one that we want to forecast. In order to clearly distinguish it from the causal variables its name will be called $Y$. There is only one such variable in any particular treatment of the data. However, it is quite possible to have different cost values for the same product: for instance you may have the cost of development, the cost of manufacturing, the cost of materials, etc. And you may be interested to find out a correlation between the development cost and the production cost. In such a case, you will have to define, for each treatment, which is the "dependent" variable, and which are the causal variables.

## The Observed Values

A value is a figure (for a quantitative variable) or an attribute (for a qualitative variable) observed on a product. Values for products belonging to the sample will always be represented by small letters, capital letters being reserved for products of the population.

---

[4] In such a case it must not be confused with a quantitative variable: we can attribute the integer "1" to the sentence "made by machine C", etc., which does not mean that such attributes may be added or multiplied.
[5] We return to this important subject – from a methodological point of view – in Volume 2, Chapter 1.

## For the Dependent Variable

As previously mentioned, there is only one dependent variable per product. The table of the values for the $I$ products therefore takes the form of a vector:

$$\begin{Vmatrix} y_1 \\ y_2 \\ \vdots \\ y_i \\ \vdots \\ y_I \end{Vmatrix}$$

or simply $\bar{y}$ (column vector). $y_i$ represents the cost of product $i$.

## Centered and Scaled Variables

The values, once centered, are noted $_c y_i$; once centered and scaled are noted $_{cs} y_i$. The definition of these values requires the knowledge of the arithmetic mean, noted $\bar{y}$ and the standard deviation, noted $s_y$ which are defined in Chapter 2 of Volume 2. Then:

$$_c y_i = y_i - \bar{y} \qquad _{cs} y_i = \frac{y_i - \bar{y}}{s_y}$$

Other notations for the sample.
Notations for the sample always use small letters:

- Arithmetic mean – or simply "mean": $\bar{y}$.
- Median: $\tilde{y}$.
- Dynamic center (the term is defined in Chapter 9 of Volume 2) in general: $\hat{y}$, and its value for product $A_i$: $\hat{y}_i$.
- When data point $i$ (one product) is eliminated from the sample: $y_{(i)}$ represents the set of all values less $y_i$.

## Residuals and Euclidian Distances

Residuals are an important concept in this book; it quantifies, for a given product, the "distance" between the dynamic center and the cost value. As several "distances" may be defined, one distinguishes $e_{i+}$ defined by $e_{i+} = y_i - \hat{y}_i$, $e_{i*}$ defined by $e_{i*} = (y_i/\hat{y}_i)$, $e_{i-}$ defined by $e_{i*} = (y_i/\hat{y}_i) - 1$.

The Euclidian distance used in a sample is represented by $\Delta_i$
A normalized residual is noted $e^*_{i+}$ (defined only for additive residuals).

## For the Independent or Causal Variables

An *observed* value for a variable is always represented by a small $x$. One can say: we measured $x = 17.5$ kg.

An observed value always refers to a variable and a product: it is the value observed for product $i$ when variable $j$ is considered. In order to make this clear, an observed value, when there is more than one causal variable, must always have two indexes. These indexes will always be in the following order: product number, variable number, both being separated by a comma. The value observed for product $i$ on variable $j$ will therefore be represented by $x_{i,j}$:

$$\text{observed value: } x_{\text{product\_number, variable\_number}}$$

Observed values are generally arranged in tables. Tables will play an important role in this book. They must therefore be fully understood.
Example:

A column is dedicated to a variable: it gives the values observed on all the products for this particular variable
⇓

A row is dedicated to a product: it gives the values of all variables observed for this particular product ⟹ $\begin{Vmatrix} 23 & 45 & \ldots & 52 \\ 47 & 15 & \ldots & 37 \\ \ldots & \ldots & \ldots & \ldots \\ 67 & 17 & \ldots & 75 \end{Vmatrix}$

Such a table – the mathematical term of "matrix" will generally be used – will be represented by a letter inside two sets of two small bars, just to remind the reader it is a special entity, such as $||x||$ for the observed values.[6]

$||x||$ is the basic matrix or set of observed values. An element of this matrix is marked by two indexes giving its row number and its column number, both numbers starting at 1. The following basic rule will be always applied:

$$\text{element of matrix } ||x||: x_{\text{row\_number, column\_number}}$$

This means that a row is dedicated to an object, a column being dedicated to a variable.

$||x||$ contains the raw data directly given by the observations (the sample). However, several matrices, derived from this raw matrix, will have to be used in the computations. They will be referred to by a pre-index or exponent:

- $||^+x||$ is derived from $||x||$ by adding a first column filled with 1. This column represents a constant value which is "observed" on all objects.
- $||_c x||$ is derived from $||x||$ by "centering" all the quantitative data. This centering proceeds column per column (centering has no meaning for the rows), each column being dealt with independently from the other ones:
  - the mean or average value of column $j$ is computed; it can be called $x_{\bullet,j}$, the little hat reminding it is an average,
  - this mean value is subtracted from each value of the column.

---

[6] Symbol X (in capital letters) is generally used in most textbooks, sometimes in bold or italic characters. If you are not very familiar with matrices computations, it makes the reading confusing. To facilitate this reading for all cost analysts, the small vertical bars were added in this book.

# Notations

- $\|_s x\|$ is derived from $\|x\|$ by "scaling" the data. This scaling proceeds column per column (scaling has no meaning at all for the rows), each column being dealt with independently from the other ones:
  - the standard deviation of column $j$ is computed; it can be called $s_{\bullet j}$,
  - each value of the column is divided by this standard deviation.
  
  Such a matrix is very rarely used, the data, before scaling being nearly always first centered.

- $\|_{cs} x\|$ is derived from $\|x\|$ by centering and then scaling all the quantitative data. In this process an element $x_{i,j}$ becomes:

$$x_{i,j} \rightarrow {}_{cs}x_{i,j} = \frac{x_{i,j} - \overline{x}_{\bullet,j}}{s_{\bullet,j}}$$

The major advantage of this process is that now all the variables have the same unit, whatever they represent (mass, energy, speed, etc.).

These pre-indexes can be used together. For instance matrix $\|{}^{+}{}_{cs}x\|$ represents the matrix derived from $\|x\|$ by adding a first column of 1, centering and scaling the quantitative data. *Note*: In the centering and scaling process, the column of 1 – as well as the qualitative variables – remains unchanged: it is not concerned by these two processes.

We may have to use matrices from which a row (a product) or a column (a variable) is deleted. The following symbols will be used:

$\|x_{[i,\bullet]}\|$ represents the matrix $\|x\|$ when row $i$ is deleted.
$\|x_{[\bullet,j]}\|$ represents the matrix $\|x\|$ when column $j$ is deleted.

## Mathematical Symbols

log or $\log_{10}$ – sometimes used for the sake of clarity – represents the logarithm in base 10.

ln represents the natural logarithm (base $e = 2.71828$).

# What You Need to Know About Matrices Algebra

You do not need to know so much …
If you want to study the subject in depth, one can recommend Draper and Smith [20] as an introduction, Lichnerowicz [36] and Golub and Van Loan [31] as full – sometimes complex – developments.

## Matrices Are First a Stenography

Matrices are first used because they are a very simple and powerful stenography: it is always easier to mention the set of values as $\|x\|$, instead of displaying the whole table of these values.

Matrices, in this book, always contain real – or ordinary – numbers. The set of all these real numbers is represented by the symbol $\Re$ (another stenography), the set of all positive numbers being noted $\Re^+$. A matrix has a size given by its number of rows, let us call it $I$, and the number of columns, let us call it $J$ (or $J + 1$ if a column of 1 is added); it therefore contains $I \times J$ elements. All matrices containing this number of elements (which are real numbers) are said belonging to the set $\Re^{I \times J}$. A particular matrix of this size is therefore said to belong to $\Re^{I \times J}$, or simply to be a matrix $\Re^{I \times J}$, $I$ being the number of lines, $J$ the number of columns.

### General Properties About Matrices

The row "rank" of a matrix is the largest number of linearly independent rows; the column "rank" is the largest number of linearly independent columns. It can be demonstrated that for any matrix both ranks are equal: so one can speak only about the **rank** of the matrix.

A square matrix is said to be "full" rank if its rank is equal to the smallest of $I$ or $J$ (as $J = I$). A square matrix (the notion of singularity applies to square matrices only) is said to be **singular** if it is not full rank; its determinant is then equal to 0.

A matrix of which determinant is close or equal to 0 is said to be "*ill conditioned*".

## Particular Matrices

A matrix such as $I = 1$ and $J = 1$ (it has just one element) is a scalar. We will consider it as an ordinary number.

A matrix such as $J = 1$ (one column then) is said to be a **column-vector**, or simply a vector; an example was given by the set of the $y_i$ (the list of the costs for a set of products). It is represented either by $\|y_i\|$ or more commonly by $\bar{y}$.

A matrix such as $I = 1$ (just one row then) is said to be a **row-vector**. An example is given by the values of all the variables observed for a particular product. If we call $z$ these values, such a row-vector is represented by:

$$\|z_0, z_1, \ldots, z_j, \ldots, z_J\| \text{ or } \bar{z}$$

Neither confuse a column-vector and a row-vector. They are different entities, as the examples given illustrate: the column-vector groups homogeneous values (costs for instance), whereas the row-vector groups inhomogeneous values (for instance: mass, power, material, etc.). The product of $\bar{y}$ by $\bar{z}$ has a meaning (if the number of elements is the same) whereas the products of $\bar{y}$ by $\bar{u}$ or $\bar{v}$ by $\bar{w}$ have no meaning at all, even if they have the same number of elements.

A matrix having the same number of rows and columns is said "**square**", or $\Re^{I \times I}$.

A **diagonal** matrix is a matrix of which all the elements are 0 except the elements in the first diagonal. The following example illustrates what the first diagonal is:

$$\begin{Vmatrix} d_1 & 0 & \ldots & 0 & \ldots & 0 \\ 0 & d_2 & \ldots & 0 & \ldots & 0 \\ \vdots & \vdots & & \vdots & & \vdots \\ 0 & 0 & \ldots & d_i & \ldots & 0 \\ \vdots & \vdots & & \vdots & & \vdots \\ 0 & 0 & \ldots & 0 & \ldots & d_I \end{Vmatrix}$$

One can also define bidiagonal or tridiagonal matrices, but you do not need them, except if you want to compute by yourself the SVD (which stands for "singular values decomposition" of a matrix).

The "*trace*" of a square matrix is the sum of the elements of its first diagonal. For the matrix just defined, the trace is equal to $\sum_i d_i$

A triangular superior matrix is a matrix of which only all the elements of the superior triangle are different from 0, as for instance the following matrix:

$$\begin{Vmatrix} 0 & 7 & 2 & 8 \\ 0 & 0 & 5 & 3 \\ 0 & 0 & 0 & 1 \\ 0 & 0 & 0 & 0 \\ 0 & 0 & 0 & 0 \end{Vmatrix}$$

A **symmetric** matrix is a matrix of which values are symmetrical in respect to the first diagonal.

An **orthogonal** matrix is a matrix of which the inverse is equal to its transpose (both terms are defined later on): $\|M\|^{-1} = \|M\|^t$.

An **idempotent** matrix is a matrix of which square (the product – the term is defined below – of the matrix by itself) is equal to it: $\|M\|^2 = \|M\| \otimes \|M\| = \|M\|$. An idempotent matrix has special properties:

- The trace of an idempotent matrix is equal to its rank.
- Its eigenvalues are only 0 and 1.

## Algebra of Matrices

We are still in the domain of stenography. Here we just need rules.

The **inverse** of $\|f\| \in \mathfrak{R}^{I \times J}$ is the matrix noted $\|f\|^{-1} \in \mathfrak{R}^{J \times I}$ (notice the dimensions) which is such that:

$$\|f\| \otimes \|f\|^{-1} = \|1\| \quad \text{with here } \|1\| \in \mathfrak{R}^{I \times I}$$

An important theorem about inverse is that the inverse of a product is given by the inverse of each matrix, the product being taken in the reverse order (this is obvious to save the rule about the matrices dimensions):

$$\left(\|f\| \otimes \|g\|\right)^{-1} = \|g\|^{-1} \otimes \|f\|^{-1}$$

The **transpose** of a matrix $\|f\| \in \mathfrak{R}^{I \times J}$ is a matrix noted $\|f\|^t \in \mathfrak{R}^{J \times I}$ obtained by interchanging the rows and columns: column 1 of $\|f\|$ becomes row 1 of $\|f\|^t$, etc.

An important theorem about transposition is that the transpose of a product is given by the transpose of each matrix, the product being in the reverse order (this is obvious to save the rule about the matrices dimensions):

$$\left(\|f\| \otimes \|g\|\right)^t = \|g\|^t \otimes \|f\|^t$$

With the rule given for transposition, the transpose a row-vector is a column-vector, and the reciprocal. This is an important application of the transposition.

## Operations on Two Matrices

Two operations can be defined on matrices: addition and multiplication. Addition will use the symbol $\oplus$, multiplication the symbol $\otimes$; these symbols are just there to recall the reader that these operations are not "ordinary" operations.

You can **add** two matrices ONLY IF they have the same $\mathfrak{R}^{I \times J}$ type (the same size). The sum of two matrices is a matrix of the same type, of which element $i,j$ is the sum of the corresponding elements of the original matrices: the operation $\|u\| \oplus \|v\|$ gives a matrix $\|w\|$ with the simple rule $w_{i,j} = u_{i,j} + v_{i,j}$. The "neutral" matrix for the addition is the matrix, noted $\|0\|$, of which all elements are equal to 0:

$$\|u\| \oplus \|0\| = \|u\|$$

if, of course $\|u\|$ and $\|0\|$ have the same type.

You can **multiply** two matrices ONLY IF the number of lines of the second matrix is equal to the number of columns of the first one: the product $\|f\| \otimes \|g\|$ *in this order* where $\|f\| \in \mathfrak{R}^{I \times K}$ and $\|g\| \in \mathfrak{R}^{K \times J}$ gives a matrix $\|h\| \in \mathfrak{R}^{I \times J}$: the mnemonic rule is that, when you write the matrices in the order you want to multiply them, the indexes which are "in the middle" (here $K$) must be equal and disappear in the operation. Note that the multiplication is not commutative ($\|g\| \otimes \|f\|$) if it is possible, is different from ($\|f\| \otimes \|g\|$) both operations are of course possible if both matrices are square and of the same size. The element $h_{i,j}$ (row $i$, column $j$) of the matrices product is given by the sum of the products, term to term, of the elements of row $i$ of matrix $\|f\|$ by the elements of column $j$ of matrix $\|g\|$:

$$h_{i,j} = f_{i,1} \cdot g_{1,j} + f_{i,2} \cdot g_{2,j} + \cdots + f_{i,k} \cdot g_{k,j} + \cdots + f_{i,K} \cdot g_{K,j} = \sum_{k=1}^{K} f_{i,k} \cdot g_{k,j}$$

Maybe it is easier to remember the rule with a graph (Figure 1).

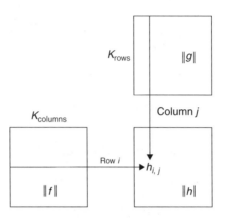

**Figure 1** Multiplying two matrices.

The "neutral" element for the product of two matrices is, for the matrix $\|f\| \in \Re^{I \times J}$, the diagonal matrix $\|1\| \in \Re^{J \times J}$. One can write, if the dimensions are respected, $\|f\| \otimes \|1\| = \|f\|$.

An important theorem about multiplication of matrices is that this operation is associative:

$$\|a\| \otimes (\|b\| \otimes \|c\|) = (\|a\| \otimes \|b\|) \otimes \|c\| = \|a\| \otimes \|b\| \otimes \|c\|$$

So, the parenthesis may be deleted.

## An Exception

A scalar was defined as a matrix belonging to $\Re^{1 \times 1}$. With this definition it is impossible to make the product of a matrix by a scalar (the dimensions rule is not satisfied). Therefore the product of a matrix $\|f\| \in \Re^{I \times J}$ by a scalar $a$ is specially defined by a matrix belonging to $\Re^{I \times J}$ where all elements of the first one are multiplied by $a$. This is a small defect in the stenography!

## Decompositions of a Matrix

A matrix can take different forms and mathematicians worked a lot about transforming forms. The advantages of these other forms are to present, in a simple way, interesting characteristics of a matrix. The most useful decompositions are the QR and the SVD.

Assume the matrix we are interested in belongs to $\mathfrak{R}^{I \times J}$.

The **QR decomposition** is the fact that any matrix can be written as the product of two matrices, the first one, noted $||Q||$, being orthogonal and the second one, noted $||R||$, being triangular superior.

The **SVD decomposition** is the fact that any matrix can be written as the product of three matrices:

1. The first one is orthogonal and belongs to $\mathfrak{R}^{I \times J}$. It is generally called $||U||$.
2. The second one is diagonal and belongs to $\mathfrak{R}^{J \times J}$. It is generally called $||D||$. It is a square matrix which has as many lines and columns as the number of parameters. The values which are in the diagonal matrix are called the "singular values" of the matrix; they are represented by the symbol $d_j$.
3. The third one is also orthogonal and belongs to $\mathfrak{R}^{J \times J}$. It is generally called $||V||^T$.

## Norm of a Matrix

There are several definitions of a matrix norm. The most frequently used is given by:

$$|M|_q = \sup_{\vec{a} \neq 0} \frac{|M \otimes \vec{a}|^q}{|\vec{a}|^q}$$

where $q = 2$ is the most common. This norm is then based on the usual norm of the vectors, defined as:

$$|\vec{a}|_2 = \left( \sum |a_i|^2 \right)^{\frac{1}{2}}$$

A quick definition of the matrix norm is the norm of the largest vector which can be obtained by applying the matrix to a *unit* vector.

## Matrices, as Mathematical "Objects", Also Have Special Properties

These properties come from looking at matrices as operators. This simply means that a matrix can be seen as an operator which transforms a vector into another vector; we write:

$$\vec{f} = ||M|| \otimes \vec{g}$$

According to the multiplication rule, if $||M|| \in \mathfrak{R}^{I \times J}$, then $\vec{g}$ must have $J$ rows (type $\mathfrak{R}^{J \times 1}$) and $\vec{f}$ will be a vector with $I$ rows (type $\mathfrak{R}^{I \times 1}$).

## Eigenvalues and Eigen Vectors of a Square Matrix $\|M\| \in \Re^{J \times J}$

Eigen vectors (also called "characteristic vectors") are vectors which are transformed by matrix $\|M\|$ into vectors parallel to themselves: if $\vec{E}$ is an eigen vector, then:

$$\|M\| \otimes \vec{E} = \lambda \cdot \vec{E}$$

where $\lambda$ is a scalar, called an eigenvalue (or a "latent root", or "characteristic root or value").

### The Case of Full Rank, Symmetric, Matrices

These are the only matrices for which we are going to search the eigen vectors and the eigenvalues.

For these matrices:

- there are $J$ different eigen vectors;
- these eigen vectors are orthogonal;
- the eigenvalues are real;
- the sum of the eigenvalues is equal to the trace of the matrix, and their product is equal to its determinant (consequently if an eigenvalue is very small, the determinant of the matrix may be small as well and its inverse may be quite large).

If $\|M\|$ is non-singular, the eigenvalues of $\|M\|^{-1}$ are the inverse $(1/\lambda_j)$ of its eigenvalues.

### Relationships with the Singular Values

The eigenvalues are the squares of the singular values:

$$\lambda_j = d_j^2$$

One advantage of the singular values on the eigenvalues is that the search for the singular values does not require the inversion of the matrix, whereas it is necessary for the search of the eigenvalues: if the matrix $\|M\|$ is ill conditioned, the latter may not exist, when the first ones always do.

# Part I
## The Framework

## Part Contents

### Chapter 1 **Cost Estimating: Definition**
This chapter gives a definition of cost, price and cost estimating, and introduces the subject by reminding the basic theorem about forecasting.

### Chapter 2 **Cost Measurement**
Cost measurement is an important subject, because no cost forecast can be better than its measurement.

It introduces the subject from an engineer's point of view, not from an accountant one. Starting from the expenditures, it goes to the definition of the expenses and briefly describes the different ways of cost measurement which are found in the industry, with the advantages and limitations, always from a cost estimator's point of view, of each of them.

### Chapter 3 **Overview of Cost Estimating**
This is an important chapter because it introduces several concepts which will be used throughout the book. It introduces first the different approaches which were found for cost estimating and then the fact that any cost estimate is based on analogy.

Then it shows that parametrics is just the modern term for speaking of quantitative analogy. However the term is mainly used nowadays when discussing about products cost estimating (which means it is very rarely used for components or activities, although it could).

### Chapter 4 **"Elementary" Cost Estimating**
This chapter just shows what can be done when the reference products, on which is based analogy, are very limited. It can be considered as a first approach to parametric cost estimating.

### Chapter 5 **An Introduction to Parametrics for the Beginner**
This chapter, using an example, shows how the "parametrician" thinks about cost estimating. Starting form some data, he/she makes some analysis on then (just for discovering potential problems) and then tries to extract what can be used for forecasting purposes; he/she always has to examine different ways for doing that until a satisfactory solution is found. Then the result of this "extraction" is used for estimating the cost of a new product.

# 1 Cost Estimating: Definition

## Summary

This chapter is just an introduction to the subject.

It starts by defining the words "cost" and "price", as they cover different concepts, even if they are so often mixed.

It then adds a few comments about the definitions of cost, as this is the subject of this book.

Our attention is then turned to what we expect from a cost estimate, which requires to define what we mean by a "good" cost estimate. A good cost estimate depends on the phase of the project it is computed for: the purpose being that it is a cost which does not distort the decision which is made on its basis. Obviously it should be close to the reality; but one thing which is very important is that it should rank all the solutions (a project always has several technical solutions to offer) with no mistake: two things are especially important during what we will call the "decision phase":

- the relative costs of the projects alternatives should be correct,
- their level should be realistic.

We rapidly then mention the three different project phases, from the cost analyst's point of view. These phases have different objectives and require different tools.

The conclusion of this chapter is what can be called the "fundamental theorem about forecasting". It will be our Ariane thread in this book.

A few comments are eventually made on the word "parametric".

## 1.1 Some Definitions

### 1.1.1 What Is Cost?

Let us start with a definition:

> Cost is the economic value of the human effort which is deployed to go, in a given environment, from situation A to situation B.

This definition requires a few comments.

## About the Economic Value

The economic value emphasizes first of all that human effort, as it is well known, is not valued everywhere the same way, second that cost has generally to be expressed in monetary terms (so many €, or so many £ or so many $, etc.). The main reason of the second point is that, although cost is always the measure of a human effort, the organization of our societies, which is partly built on the division of labor, implies that we have quite often to acquire the result of efforts made by other people (such as energy, raw material, machines, depreciation, etc.); this acquisition is made at a given price. Using the monetary values allows to easily add together efforts coming from different sources.

It may happen, in some industries, that cost is expressed in terms of hours: this is particularly true for the software industry[1] or for estimating the cost of activities such as project management. This is done for three reasons:

1. Direct human effort, as opposed to indirect human effort, is the major cost element in the software industry.
2. Putting a monetary value on such an effort is easy and has been done for years: computing the monetary value of a man × hour is one of the basic function of cost accounting, or, as we call it, cost measurement. It is then easy to go from estimated hours to cost.
3. Using hours eliminates the problem of the currency value: it easily allows to compare human efforts for projects made at different times, whereas the monetary value has to use a unit which changes with time (this change is generally, and improperly[2], called "inflation").

Another example will be shown with Peter Korda's "Value Engineering": VE computes the cost for manufacturing goods in terms of hours (see Chapter 11 of this book). The reason is that all small companies in the mechanical industry estimate first the hours needed to accomplish some tasks[3]. The "VE" tool fits to this practice.

The economic value[4] changes in space and time: a human effort had an economic value much lower in the past than it has today, and it has not the same economic value, at a given time, everywhere in the world. The consequence is that it can make comparisons of costs – from one period to another one, or from one country to another one – difficult. We will return to this subject in Part II.

---

[1] Barry W. Boehm, in Ref. [8], p. 61, writes: "COCOMO avoids estimating labor costs in dollars because of the large variations between organizations in what is included in labor costs (unburdened by overhead costs? burdened? including pension plans? office rental? profits?), and because man-months are a more stable quantity than dollars, given current inflation rates and international money fluctuations".

[2] Inflation, for the economist, means the growth of the supply of money. It was noted in the past (one of the first experiences came when the Spanish brought back from the new world a large quantity of gold and silver) that inflation and a general increase in the price of goods generally go together. From this remark, people took the habit to use the same word for the growth of the supply of money and the general growth of the cost of goods on the market.

[3] We will see that the level of effort necessary to accomplish a task is generally expressed in time units. A good example is given in Ref. [3].

[4] We should not go too far in the definitions ... If you want to go one step down, the economic value of something could be defined as its exchange value.

## About the Environment

The environment includes:

- on one hand what is available,
- on the other hand the constraints.

As the reader may expect, the definition of the human effort environment is very important to understand cost: this environment includes many things, from the education and the training of the people, the machines they use, the plant organization, the management efforts, ... up to the motivation of both management and operators. It helps explain, in addition to cost accounting practices, that two companies manufacturing the same product will never observe exactly the same cost.

The environment also includes the constraints, either legal (such as the time people are allowed to work per week, or the social benefits), or imposed by the situation (such as the project duration, or the limited choice of suppliers, etc.).

## About the Situations

Situation A refers to the situation when the effort starts; situation B is the result of this effort. In this book it will be a product, or a software, or a building, or a tunnel, ... the word "product" being used as a generic term.

A cost estimate is a prediction of what the human effort for going from situation A to situation B (the cost), in our given environment, will be.

Before we go any further it must be emphasized that the three terms of the definition should be as precisely as possible defined before any estimate can start:

- The starting point must be clearly defined. This seems obvious in many situations (we start from the "customer's requirements"!) but a close analysis shows that, quite often, the starting point is not so clear. This is generally the case when some development[5] has to be carried out. Why? It comes from the fact that development costs are very dependent on the **history** of the company: Do we have to realize something we never did before? Have we already produced something similar? Very similar or a bit similar? Do we have some – which one? – experience about the technologies that will be needed? etc. It can be said, and there is some truth about this statement, that this history is taken "unconsciously" into account; nevertheless experience shows that it is always better to properly document the "starting point" and this is specially true if we want to analyze the discrepancies between what the cost really is and the cost which was estimated. From our experience, when some development occurs, about 50% of

---

[5] Different organizations may use different words. When we are not sure on how to fulfill the customer's requirements, some studies must be carried out in order to properly define how these requirements can be satisfied. We use here the word "development" for these studies.

the differences between estimated cost and actual cost can be explained by an incorrect perception of the starting point.
- The finishing point is, generally speaking, better defined, at least when the acceptance tests that the customer will carry out (in terms of technical specifications and quality level) are clearly established before the project starts. Is that always the case? Even when it is the case, how many projects are carried out with no modification? If we want to do the analysis between estimated and actual costs, we will have to refer to what was the definition of the finishing point when the estimate was prepared.
- The environment has also to be clearly perceived. The term includes many things, from the industrial development to the availability of resources, plus, of course the constraints. One of the major constraints in many projects is the delivery date.

### 1.1.2 About Price

Here is the definition of price:

> Price is the result of a monetEary agreement between two parties when something is exchanged between them.

From these definitions, it clearly appears that cost and price are completely different concepts.

However as a company buys everything it needs (including labor) from the market, strictly speaking a cost value is only an addition of prices ... Where is then the difference? Quite obviously if a company buys an equipment to integrate it directly to the project it manufactures, the price of this equipment becomes a, direct, cost of the project. The difference comes from several points:

1. A company is more or less able, or powerful, to negotiate for the acquisition of the same good from the same market. Furthermore prices follow, more or less depending on what is priced, the basic law of "supply and demand": prices are not fix on the market place.
2. Even if it buys everything at the same unit price, a company completely controls:
   - the level or the quality of what it needs,
   - the quantities it needs for doing something,
   - the type of machines it will use,
   - the efficiency of the manufacturing process, etc.
3. Plus of course the level of automation of the process, the plant organization, etc.

Consequently if it is true that cost is only an addition of prices, the relationship between resources acquisition prices and costs of the product is not as strong as it could be expected.

Quite obviously the seller cannot ignore its cost when he/she proposes a price ...

Furthermore modern project management attempts to reduce the gap between cost and price.

## 1.2 What Is Cost Estimating?

A cost estimate is a forecast of what the cost, the use of resources, will be or should be. The distinction may be important for the decision-maker:

- The "will be cost" is a cost forecast of the *product* as envisioned by our engineers.
- The "should cost" is a cost forecast of the *function* fulfilled by the product, as the consumer sees it.

The methods presented in this book may be used for estimating one or the other cost. The distinction is irrelevant to the cost estimating methods.

Forecasting is not an exact science. Consequently any forecast should go with:

1. the set of hypotheses that define the scope of the estimate;
2. its level of uncertainty (this subject will be discussed).

Some people make a distinction between "evaluation" or "estimate" or something else; generally this distinction refers to the level of uncertainty. Being only concerned by the methods used for obtaining a forecast, we do not make any such distinction: for us they are, in this book, considered as synonymous.

### What Is a "Good" Estimate?

Answering this question is more complex that it seems. It immediately rises another question: what is the purpose of an estimate?

Estimates are needed in different circumstances; these circumstances may dictate the way the estimates are done. It is necessary here to distinguish the phases of "project" management.

These phases are rather well defined nowadays. For instance, the French Society AFITEP (Association des Ingénieurs et Techniciens en Estimation, Planification et Projets) recognizes (Ref. [1], p. 14) four levels:

- Preliminary study:
  - identification, which needs an estimate at ±45% and requires half a day;
  - feasibility, which needs an estimate at ±30% and requires 3 days.
- Realization:
  - design, which needs an estimate at ±15% and requires 15 days;
  - development, which needs an estimate at less than ±5% and requires 100 days.

The American Association of Cost Engineers (AACE) proposes five classes:

- Class 5 corresponds to a low level of project definition (2%), the purpose being screening or feasibility studies of several projects.
- Class 4 corresponds to concept study.
- Class 3 to budget authorization (level of project definition goes from 10% to 40%).
- Class 2 to bid tender.
- Class 1 to check estimate (level of project definition goes from 1% to 4%).

Costs are not really estimated by the cost estimators, but by the managers. Managers, due to their experience, the knowledge of the market, the comparison with other products, their commercial strategy, etc., have a preconceived idea of what the cost "should be", but ask their cost estimators to propose a figure. When these persons come with such a figure (the "will be" cost), you can be pretty sure that this cost will be too high; it comes from the fact that the managers think in terms of functions to be performed (plus the constraints of the market or the available budget), whereas costs estimators think in terms of products. Then the designers and the cost estimators redo their work in order to see how the "should cost" could be obtained. The real interest of cost estimating is not only in the forecast of the "will be" cost, which is nevertheless a good starting point, but also on the way the functions may be achieved for the "should cost".

This idea is nowadays very popular with the concepts of "design to cost", or the cost as an independent variable (CAIV) which is an extension of the previous one. Many books have been written on the subject which implies a close cooperation between the designers and the cost estimators. As this book is made for the cost estimators, it is mainly dedicated to the "will be" cost; in other words it tries to answer the question: "Describe to me what you want to do (+ the constraints) and I will tell you, by careful extrapolation of the results of past experiences, which I recorded for years, what it will cost".

In this book, for the sake of simplicity on one hand, for the objective of the cost estimates on the other hand, we will recognize in the project life cycle only three major phases, each one creating new challenges to the cost estimator. Our focus is not here on the accuracy of the estimate, but rather on the purpose of the estimates and, as a consequence, on the way they can be achieved. They are called here:

1. the "decision" phase,
2. the "validation" phase,
3. the "execution" phase.

These phases are completely different in their definition and the way cost estimates are dealt with:

1. The *decision phase* (which includes about the first three levels of AFITEP) corresponds to the period of time during which major decisions have to be made about the project, from "Should we prepare a proposal to this 'request for proposal' (RFP)?", up to "here is the way we can technically respond and the price of our quotation will be about this amount!". The work of the cost estimator is there to quickly assist the decision-maker: estimating the cost of various concepts, estimating the influence of technical uncertainties on the cost, etc. The focus during this whole phase is the *product*; the estimate becomes, as time goes on, more and more detailed, as we go down on the product tree.
2. The purpose of the *validation phase* is completely different: the manager has decided the way we will technically respond and the price for which we have a good chance to be competitive. Now the work of the cost estimator becomes to help the project manager to answer the question: "How can we make sure we will accomplish the project for the cost which has been decided upon?". The focus, which is always to help the decision-maker (who is now the project manager), now shifts to *activities*, because activities are the means from which the project will be accomplished. The project manager thinks about several ways of making the project a reality and the cost estimator must estimate, in more and more details, the cost of each, until a solution is found which fits inside the budget, the technical objectives (including the quality level) and the duration.

3. During the *execution phase*, the role of the cost estimator changes again. The project goes on and accumulates a lot of information. What has to do the cost estimator? He/she must now, periodically, from the information which is collected, decide if the project will, at completion, remain inside the allocated budget; he/she does that by extrapolating the cost and produces the estimate at completion (EAC).

These phases correspond to a reality: when you hear the discussions between engineers during the decision phase, you realize that they turn around the product; during the validation phase, they turn around the activities; until finally the constant question becomes: "Shall we fit, at completion, inside the allocated budget?". Quite obviously the work of the cost estimator will change during these three phases: their objectives are different and, consequently, he/she needs different tools.

Barry W. Boehm writes (Ref. [8], p. 32) "a software cost estimating model is doing well if it can estimate software development costs within 20% of the actual costs, 70% of the time, and on its own home turf (that is within the class of projects to which it is calibrated)". And he adds: "this is not as precise as we might like, but it is accurate enough to provide a good deal of help in software engineering economics analysis and decision-making". The focus is obviously on decision-making.

A good forecast is a prevision which is attainable and therefore helps the decision-maker to envision the future and consequently manage the business.

A good estimate is then an information the decision-maker may rely on when he/she makes a decision. It is generally not the figure that will appear from cost accounting at the end of the project. This immediately arises a question the cost estimators are very sensitive about: Must we compare cost estimates and observed (measured) cost?

It is a question some cost estimators refuse to answer to. Their argument is that between the time they made an estimate and the time true costs are known, many things happened: there were some changes in the products, some unforeseen difficulties arose, the environment on which the estimate was based had changed, the constraints are different from what was considered, etc.

They are obviously right. Nevertheless the comparison is legitimate for at least two fundamental reasons:

1. Credibility is a major element in cost estimating, and in any type of forecast anyway (think about the way you listen to meteorological previsions for instance). The credibility a manager gives to a cost forecast is based of course on the known experience of the cost estimator, or the lack of experience if a completely new subject is considered, and the seriousness the cost estimator does his/her job (assuming that the manager has formalized the periodic audit of cost estimating, as it will be mentioned in Part IV). It is also built on the comparisons between forecasts and actual costs. A manager will not have the same attitude toward cost estimates if the deviations between previsions and reality are at $\pm 40\%$ or $\pm 5\%$! Add to that a manager is perfectly able to take into account when building his/her credibility the modifications which occurred in the project, if these modifications are clearly documented (this is a good reason for the cost estimator, when he/she prepares an estimate, to list the hypotheses [even if we know

that hypotheses are generally quickly forgotten …] on which the estimates are based).
2. In any science, the only way we can, as human beings, improve our predictive capabilities is to compare forecasts and reality: there is no other way. Not doing so is the right path for sclerosis. We recognize that it is not always pleasant to admit a large deviation between both, but looking for explanations, beyond the modifications and the changes in the environment, is the only way we can improve ourselves.

## 1.3 The Fundamental Theorem About Forecasting

The fundamental "theorem" for forecasting in general and how this theorem applies to cost prediction, whatever the way it is prepared, is the following one:

In any science, we can only forecast by extrapolating from the past.

This does not mean we just look at the trends. It means we have to extract from past observations some concepts, some relationships, some guidelines, from which the future may, within a confidence interval, be predicted.

To the trend you may have to make corrections if the technology, the organization, the materials, etc… have changed. What we mean is that in order to predict the future you always start from the past (let's call it "the reference") and adjust it for taking into account what has changed. The important point is that you need a starting point.

Two very important points are in the previous comment: a cost estimate is obviously a figure, but - and this is often forgotten - should also include an information about its credibility; this information may take different forms, the most frequent being the confidence interval of the estimate.

### Consequences

The first and obvious consequence of the fundamental theorem is the necessity of creating a database. You must not be afraid by this word: a database is not necessary a giant list of data. It can be limited and sometimes very limited; in such a case you will need more complex tool for the extrapolation study, but it must exist.

The second point is that you need tools to carry out the extrapolation. Extrapolating is not, generally, a simple linear projection of past observations. It can be rather complex and we will see that before, it can be done, a data analysis must be carried out.

The quality of your estimate will be directly linked to the judgment you make about the quality of the data from which you extrapolate on one hand, on the choice and the quality of the tools you use on the other hand. Both things must go together: no extrapolating tool can compensate lack of data or simply poor data.

Let's repeat what was said in the preface: cost estimating is based at 30% on the data, at 30% on the tools and at 40% on judgment and a minimum of 80% must be attained in order to prepare a realistic estimate.

## 1.4  A Few Comments

### A Distinction

Cost estimating requires different people with different interests and different levels of experience. Shall we call them "cost analyst" or "cost estimator"?

In this book the *cost analyst* is defined as the person who builds, or buys and then adjusts, cost estimating tools; *cost estimator* is the person who uses these tools for making a cost estimate.

Quite often the same person is both analyst and estimator. But this is not necessary, especially in large companies.

### About the Word "Parametric"

Be careful about the word "parametric". This word may have different meanings for different persons (and can therefore be misleading):

- For the mathematician, a parameter means an auxiliary variable which slightly changes the curve, or the function, he/she studies.
- For the statistician, it means that the distribution he/she is interested in has a known shape and that this shape if fully described by two or more "parameters". For instance he/she may say that the observations fit well a "normal" (we should say (because other distributions are not abnormal!) a "Lagrange–Gauss") distribution of which parameters are the mean and the standard deviation.

  When a statistician speaks about non-parametric relationships, he/she means that he/she does not have to refer to any distribution: the relationships do not depend on a specific distribution. A good example is given by studies on the ranks.

  We will often refer to such non parametric relationships (and we will sometimes say that these relationships should be preferred to "parametric relationships"). This may sound strange in this book … but it only means that the same word has different meanings!
- For some authors (for instance Ratkowsky in Ref. [47]), the parameters of a relationship are what we call the coefficients ($b_0, b_1, b_2, \ldots$). This usage is similar to the statistician vocabulary.
- For the cost analyst a parameter is quite often a "cost driver" (or a "causal" variable): it is a physical property, which may be quantitative (such as the size) or qualitative (such as the quality level), or even subjective (such as an opinion on the complexity of the product), which, he/she thinks, has an influence on the cost.
- Let us give another example about the confusion brought by the word "parametric". Some cost estimators say that they use parametric cost estimating because they use the hourly rate as the parameter. From the mathematician's point of view they are right; but it is not the definition we use in this book. In the following chapters a parameter refers to the description of the product (e.g. the weight, the number of parts, the tolerances, etc.) or of the manufacturing process (such as the type of machine which can be used, the level of automation, etc.).
- There is therefore some confusion about the vocabulary!

We do not like so much a word which has different meanings for different persons. We used it in this introductory chapter, because it is well known in the community of people who use explicit cost drivers to estimate cost.

We will however mainly use the word "parametric" as an adjective for referring to the basic cost estimating method. We will rarely use the word "parameter", preferring the word "cost driver" or simply "variable", a variable being a concept which is used in the relationships.

The word "coefficient" will be used for the factor, or the exponent or anything else, related to a causal variable.

# 2 Cost Measurement

## Summary

Cost measurement is not a standard term. Accountants, who are the experts in this domain, call it "cost accounting". But this word does not speak so much outside their world. What we want to do is to put a cost value on all the "objects" we manufacture, from observations of the efforts devoted to the manufacturing process. Any technician or engineer knows what measurement means: it is part of their day-to-day activity. The same thing has to be done for costs and it is the reason we use the word "cost measurement".

No engineer will use a figure resulting from measurement without being able to make a judgment on its value: does this figure correctly reflect the reality? The situation is similar for the cost analyst, even if the conditional tense has still to be used; no cost analyst should use a cost value without asking the question: does this cost value effectively represent the effort which was spent to manufacture this product?

This implies first that the cost analyst knows how costs are measured. Unfortunately there are different ways to measure a cost, and the results of these different ways cannot be compared, up to the point they may be completely different.

The cost analyst should therefore understand these different ways and draw consequences of this knowledge in order not to make mistakes.

He/she should also not forget that behind the theory of cost measurement there are human beings who can make mistakes and misinterpretations of the rules, or, in the worst cases, have an interest in distorting cost values when it is possible.

Parametric cost estimating is a powerful tool for checking the reliability of cost figures, just because its main objective is to compare costs and consequently it may reveal inconsistencies. In some organizations, one of the first purposes of using parametric cost estimating is precisely to "clean" the databases.

This question is asked in a completely different way when prices are compared. It is much more difficult, but possible.

Cost estimating, as we mentioned it in the previous chapter, is based (whatever the technique used to prepare it) on a comparison with previous experiences.

The first thing to do is then to accumulate cost information from different objects or situations. The problem with cost, which has no comparison with other subjects, is that there is no way to directly "measure" cost; cost measurement is always an indirect process, sometimes extremely indirect.

What we propose to do in this chapter is to present first the various methods for cost measurement, and then to discuss about the reliability of the obtained figures and the level of confidence we may have in them.

First of all, we want to remind the reader on the way costs are "measured". This is important for two reasons:

1. The cost analyst must know there are various ways to measure cost; when he/she receives figures, he/she must be sure that the figures were computed about the same way. Of course, if these figures are prices, he/she will assume that the figures are "full cost + margin" (the margin can be positive, negative or zero).
2. Understanding the process of cost measurement allows the cost analyst to understand the limitations of the process. One of these limitations is the accuracy he/she might expect from the costs figures he/she gets from his/her own organization.

This chapter was not written to make the reader a specialist on cost accounting; cost accounting is a profession of which techniques are sometimes rather complex. Its purpose is to have general understanding of it in order to be able to read other books or manuals without falling into any excess which could not help him/her for cost estimating.

Text books (Anthony [4]) generally distinguish three accounting:

1. *Financial accounting* of which purpose is to produce financial statement that convey information to outside parties. This accounting is governed by "generally accepted principles", the purpose of this governance being that outside parties want all the companies to present their information in the same way.
2. *Management accounting* which provides information useful for the operation of the company. There may be differences with the previous in the depreciation cost (this world is defined downwards), in the value of the inventory cost, etc.
3. *Cost accounting* is the technical process by which expenses are allocated to products. It serves, with some differences, both previous accountings.

Cost measurement is not a gadget: it has been said (Elphen, etc.) that "management accounting practices are one of the major reasons for the deterioration of the US productivity".

## *The Purposes of Cost Measurement*

Cost accounting has three functions:

- *External*: for reporting to the outside world.
- *Internal*:
  - determination of the cost of the goods produced;
  - determination of the rates (labor rates, machine rates, etc.).

The second and third ones directly interest the cost analyst.

Variable costing is only for budget preparation: when preparing a budget, the load of the company is still uncertain which implies that the expenses cannot be determined with precision. One of the techniques which can be used is to compute several budgets, depending on various load. This technique does not concern the cost analyst, and will therefore not be discussed.

Unfortunately cost accounting is not made for objects or products cost measurement but first for external reports. What we mean by that is that the cost estimator needs a cost information much more detailed than the financial people do: for instance if several prototypes are built, he/she would like to know the cost of each prototype; he/she would like to know the cost of making the drawings (plus of course the number of drawings); etc. Whereas the financial people just want to know if the development cost is under budget.

## 2.1 The General Framework of Cost Measurement

We already mentioned that there are several ways to measure costs. Fortunately enough, these ways have common parts: more exactly each way is a "refinement" of the previous way. It is therefore possible to introduce these different ways by presenting them according to a tree structure (see Figure 2.1).

The starting points:

1. *Everything starts from expenditures*: an expenditure is an amount of money paid for acquiring an asset or a service. An expenditure is always recorded on a piece of paper.
2. *A major constraint in cost measurement is the accounting period*: the functions carried out by the company must be accomplished during a given period, let us say for a year, generally starting on January 1st, finishing on December 31st (but other dates are sometimes used). This constraint is understandable: external parties logically ask for periodic statements.
3. *Companies generally produce several (sometimes a lot of) different products*: each product consumes an amount of resources which may be different from the other one; the problem is then to put a value on all the resources consumed by each product.

### From Expenditures to Expenses

Expenses are the amounts of money which are used during a given year (the accounting period) for the production of the goods and services sold by the company. Going from expenditures to expenses is not a simple problem: Anthony [4] said that "the problem of measuring expenses and revenues in an accounting period is the most difficult problem in accounting". Consequently we will limit the discussion here to a simple presentation.

Not all expenditures are expenses:

- First of all a few corrections are made, such as the pre-paid expenses (expenditures related to another year) or the deferred charges. These adjustments are small and do not really concern us: it is the job of the accountant to make these corrections.
- Then, and this is very important for us, long-lived assets cannot be considered in full as expenses for the considered accounting period. They deserve a special treatment which must be known by the cost analyst: this treatment is called "depreciation".

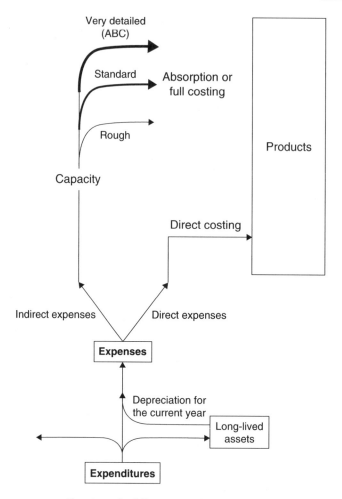

**Figure 2.1** The different ways of measuring cost.

From a cost analyst's point of view, depreciation is simply the fact that – and the major constraint mentioned earlier finds here its application – the expenditures for the acquisition of long-lived assets (such as buildings, machines, etc.) should certainly not be considered as expenses for the coming accounting period: their life time is longer – and sometimes much longer – than the accounting period. It seems logic to "spread" these expenditures on several accounting periods.

Depreciation is the result of this spreading: for each accounting period only a part of the expenditures spent for the acquisition of the long-lived assets is considered as an expense. One could also say that these assets are transferred to the next accounting period at a price lower than their purchase price: the assets which will be used for this accounting period are "depreciated" compared to their purchase price. And of course this is also true for the following period: each time an asset is transmitted to the next accounting period, its value is decreased or depreciated by a certain amount, until this value reaches 0; in such a case the asset is said to be "fully depreciated" (this does not mean it cannot be used anymore, as the financial depreciation is not necessarily equal to the physical deterioration of the asset).

Some people present the depreciation as an amount of money which is set aside from the revenue of the accounting period in order to buy the same asset when it will be worn out, whereas we present it here as the spreading of an expense. Obviously the two presentations are the two faces of the same coin. We prefer the second one when the presentation is made to people dealing with costs, not with revenues.

How are the depreciation rates computed (the term is used in plural, because each asset has its own rate)? We find here an important distinction between financial accounting and management accounting:

- Management generally prefers, because this is the image of the reality and therefore does not distort the picture, to depreciate assets at the same time they physically deteriorate.
- Financial people prefer to depreciate these assets as fast as possible, one of the reasons being to decrease the amount of the taxes.

If cost management may apply its own rules on depreciation, third parties require that the same rules be applied about the same way to all companies: this is the consequence of the "generally accepted principles". These principles help define the duration of the depreciation period on one hand, the speed of it on the second hand (several rules can be applied, the linear being a common example). This duration goes from infinite (grounds are not depreciated at all) up to about 2 years for equipments which are quickly worn out or obsolete.

At the end of the process, only the expenses remain, the purpose being now to distribute all of them on all the products manufactured by the company.

Also note at this stage that general and administrative (G&A) expenses (headquarters, interest on borrowed money, administration, etc.) are excluded from the costs which will be allocated to the products. The idea is to compute the "pure" production costs. In order to recover these expenses, they are added, for pricing purposes, to the production costs as a percentage called G&A expenses; profit follows the same logic. At the end of the process, the selling price is determined, when it is possible to do so.

## 2.2 Two First Distinctions: Fix and Variable, Direct and Indirect

We introduce in this section the term of "cost" which is the concern of this book: *an expense becomes a cost as soon as it is allocated to something* (product, project, activity, capacity, etc.). As we suppose that expenses are now going to be allocated, we will use the term "cost" from now on.

### Fix and Variable Cost

The distinction between fix and variable costs, which appears in all text books, is not very important for the cost analyst. It is just mentioned here because it is a familiar distinction. A cost is considered as fix if it does not depend on the production quantity: the depreciation of the building is the most known example of fix cost.

A cost is considered as variable if it depends on the production quantity. Examples: raw material, energy, labor, etc.

The distinction has a limited interest for two reasons:

1. Most of the costs are at the same time fix and variable: the maintenance of a machine for instance has to be done even if the machine does not work, but the level of maintenance grows with the production quantity. We will return to this subject in Chapter 9 where we examine how the costs do change, when the production load changes.
2. Very often it is the result of a management decision: the depreciation of a machine may be considered as fix if it based on the accounting period, or variable if it is computed on the level of activity.

The primary interest of the fix and variable cost is related to the preparation of the budgets: as the level of activity is not known yet, it is useful to consider (even if it is not exactly correct) that there are fix costs and variables ones, the latter depending on the activity level. This is one of the basis of flexible budgets.

## Direct and Indirect Cost

This distinction between direct and indirect costs is the most important one; it exists in all accounting systems.

Both the terms are related either to products (or projects) or to the plant itself:

1. A cost is considered as a *direct cost* (of a project), if it would disappear if this project was cancelled. A direct cost is consumed for just one project. Examples: raw material bought for this project, assets (machines, tools or sometimes even buildings) which were bought for this project only, personnel who work only for this project, etc.
2. An *indirect cost* is consumed by several projects; it is first allocated to the plant. Most often it is the most important part of the cost, from the buildings to the maintenance, the janitors, the cafeteria, the management, etc. Indirect costs are spent to operate the plant, independently of the products it manufactures. For this reason they are often called the "capacity cost". Other terms are "overhead" or "burden".

The distinction can go very far. For instance the bonus given to a salesman is considered as indirect if it is computed on the total sales, as direct if it is computed for each product sold (it may come from a management decision, or to the fact that the rate may not be the same for each product).

## About Direct Costs

Official documents (for instance Ref. [19]) distinguish five different direct costs:

1. *Materials*: the manual indicates that materials should have three general characteristics to be classified as direct costs, namely (a) they should become part of the product, (b) quantity and price should be significant and (c) measurement of quantity and cost should be relatively easy and inexpensive.
2. *Factory labor*: fabrication, assembly and quality control. The manual indicates that some categories of labor (engineering, tooling, planning, estimating, expediting) are sometimes considered as direct, sometimes as indirect.

3. *Engineering labor*: design engineering, manufacturing engineering, quality assurance, reliability, maintainability.
4. *Tooling*.
5. *Other direct costs*: special insurance and travel expenses, packaging, transportation, consultant's fees, etc.

## About Indirect Costs

The same document distinguishes the following:

1. *Material overhead*: acquisition, transportation, receiving, inspection, handling and storage.
2. *Engineering overhead*.
3. *Manufacturing overhead*: foremen, superintendents, maintenance, stockroom personnel.
4. *G&A expenses*.

A word of caution: many people confuse "direct" and "variable" costs. It is true that many direct costs are variable costs, the common examples being the labor and the material. However, terms are not synonymous: a direct cost may very well be fix (an example is given by the depreciation of a machine specially bought for a project) and an indirect cost may very well be variable, as the example of machine maintenance illustrated.

## 2.3  Direct Costing

Direct costing is a cost measurement process which stops the allocation process here:

- Direct costs are allocated to products, or services.
- Indirect costs are globally allocated to what is called the "plant capacity".

This means that the "costs" are far below the prices you may have to pay to buy the goods! Some companies really manage their costs this way. The process is illustrated on Figure 2.2. Expenses are allocated either to products when it can immediately be done, or to an account called "plant capacity". This account is the cost of keeping the plant running, not taking into account what it produces.

When a sale is made, the revenue is allocated to the product for which the sale is made. At the end of the accounting period, the difference, for each product, between the revenues and the (direct) costs is called the "contribution" of the product, contribution to the capacity cost of the company. The difference between the contributions of all products and the whole capacity cost is obviously the profit. Therefore, products can be listed according to their contribution and decisions may be made on keeping them or not.

The major advantage of this process is its simplicity as no (sometimes difficult and requiring many decisions as we will see it) allocation of the capacity cost to individual products.

The second advantage is to make easy several short-term decisions: suppose the plant has, at a certain time and for a limited time, some idle capacity. Management will look for some activity which could reduce it. At what price should it sell it? The

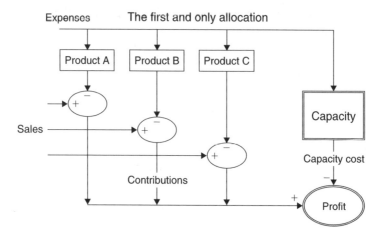

**Figure 2.2** Management by direct costing.

direct cost offers immediately a floor price: as long as the selling price is higher than this floor price, this new activity will bring some contribution to the capacity cost. Even if not important it is always better than nothing ... Of course, this is a pure short-time decision: this operation could not last very long as the full capacity cost must be, sooner or later, recovered.

Direct costing is largely spread in the businesses: as a matter of fact most of the products you buy everyday are relevant to this process. Each time the price of a product is dictated by the market, which means that the seller is small and the level of competition high, there is no other choice. When you buy your apples, it would be foolish for the retailer to compute the full cost of the sale: it would mean a lot of paper work of which utility is not obvious. It is much easier, and largely sufficient, to measure the contribution of each product to the cost of running the business (its capacity cost) and to favor the products for which the contribution is higher.

Direct costing in a company makes a difficult task for the cost analyst who would like to introduce parametric cost estimating. The problem comes from the fact that direct costs may be very difficult to compare between products, as the amount of these direct costs is often a management decision: if, for instance, a task is subcontracted to another company it becomes a direct cost of the related product. If it is made internally it does not appear in the direct cost of the product.

The conclusion is that direct costing does not really allow for parametric cost estimating.

## 2.4 Full or Absorption Costing[1]

When prices are not dictated by the market, or when a new product is considered, direct costing is not sufficient, as it is not in any case sufficient for making long-term

---

[1] The "bible" of full – or absorption-costing is Ref. [41]. Unfortunately this book confuses variable and direct costing.

decisions. Another solution has to be found. Obviously this solution implies that the capacity cost should be allocated to all products according to the resources they actually consume. It is called "full" or "absorption" costing.

As you may imagine it requires a lot of work ...

There are several ways to practice the full costing. We call them (the adjectives are not official and are only use to distinguish them):

- "rough" full costing,
- "standard" full costing,
- "detailed" full costing, also called the ABC (activity-based costing) method.

In this section we only introduce the first two ones; the last one, which is rather special (which means it is not as much practiced as the first ones) will be the subject of the following section. In these first two ways, the amount of the capacity cost on the products is called the "overhead". Before discussing the subject, it is interesting to have a quick look on what is found in the capacity cost: Figure 2.3 (Ref. [19]) illustrates. It shows the variety of costs included in the manufacturing overhead and the computation of the overhead rate, based, in this example, on the direct manufacturing cost.

## 2.4.1 "Rough" Full Costing

The procedure of allocation of the capacity cost is very simple: a rule is defined and the total capacity cost is simply allocated to the products according to this rule.

As the overheads are nearly always computed in three categories as mentioned above (material, manufacturing, engineering), indirect expenses are distributed on the products according these categories: a ratio is computed for each one and applied on the direct expenses. G&A expenses are dealt with slightly differently as illustrated on Figure 2.4.

What can be the rule? The rule depends, of course, on the type of overhead (material, manufacturing, etc.). Let us just take the example of the manufacturing overhead; several rules can be used: capacity cost can be allocated according to the following:

- *The units of production*, easy to know.
- *The materials cost*. This rule can be used if the materials are an important part of the cost of the finished products; it means that the added value is small compared to materials costs.
- *The direct labor* (hours or cost), in an opposite situation. It is true that, at least as a first approximation, direct labor could be a good indicator of the internal resources consumed by a product.
- *The direct costs* of the products, the logic being that each product consumes internal resources (procurement, manufacturing, quality control, etc.) as a function of its direct cost (which is mainly labor and material). This rule is easy to implement because the direct costs are already known.
- *The machine hours* if the manufacturing is highly automated for instance.

How does this rule apply? Suppose we observed last year, if the direct cost of the products is used as the base for allocating manufacturing expenses, a direct cost of

| | |
|---|---:|
| Salaries and wages: | |
| Indirect labor | $1,338,330 |
| Additional compensation | 80,302 |
| Overtime premium | 13,214 |
| Sick leave | 65,575 |
| Holidays | 79,164 |
| Suggestion awards | 310 |
| Vacations | 140,272 |
| Personnel expense: | |
| Compensation insurance | $ 25,545 |
| Unemployment insurance | 50,135 |
| FICA tax | 70,493 |
| Group insurance | 153,755 |
| Travel expense | 11,393 |
| Dues and subscriptions | 175 |
| Recruiting and relocation – new employees | 897 |
| Relocation – transferees | 4,290 |
| Employees pension fund: | |
| Salary | 25,174 |
| Hourly | 62,321 |
| Training, conferences and technical meetings | 418 |
| Educational loans and scholarships | 400 |
| Supplies and services: | |
| General operating | 495,059 |
| Maintenance | 9,102 |
| Stationery, printing and office supplies | 23,052 |
| Material O/H on supplies | 56,566 |
| Maintenance | 9,063 |
| Rearranging | 418 |
| Other | 3,314 |
| Hear, light and power | 470,946 |
| Telephone | 32,382 |
| Fixed charges: | |
| Depreciation | 187,118 |
| Equipment rental | 7,633 |
| Total manufacturing expense (A) | $3,416,816 |
| Total manufacturing direct labor dollars (B) | $1,340,887 |
| Manufacturing overhead rate (A) ÷ (B) | 254.82% |

**Figure 2.3** Illustration of the content of the capacity cost (manufacturing overhead). *Source*: Ref. [19].

all the products of €200 000 and a manufacturing overhead of €720 000; this means that the manufacturing overhead rate is 360%. Consequently if we get this year a new product having a direct cost of €10 000, we will conclude that the manufacturing cost is equal to 10 000 + 36 000 = €46 000.

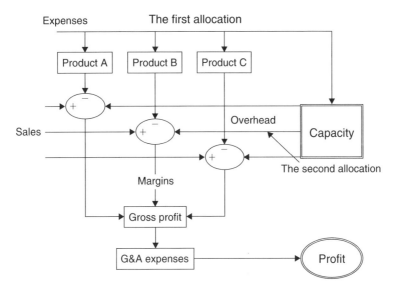

**Figure 2.4** The "rough" full costing.

The reader has certainly observed that the overhead rate applied this year is given by the overhead rate observed last year; this may introduce some distortion which would not be too serious unless:

- The basis for computing the overhead (which means the total direct cost of all the products) may change from one year to the other year.
- The value of the manufacturing overhead changes, due for instance to a change in the plant capacity or to a modification of the process. In such a case, it should not be to difficult to adjust the overhead rate computed last year before applying it to products manufactured this year.

The basic assumptions of this type of full costing are that the products are manufactured about the same way (for instance the level of subcontracting is about the same, no special machine was bought for a given product, ...) and of course that the level of internal resources is about proportional to either the direct cost or the cost of labor. This was probably the case long time ago when the capacity cost was only a small amount of the total costs (and, even if it was not fully correct, the level of imprecision was small): this explains the origin of this method. But nowadays, due to the amount of automation and robotization, the depreciation costs are a very important part of the capacity cost. It means that the allocation of the capacity cost may multiply by four or five the direct costs[2].

It might still be a solution in small businesses, where, if the products differ, the way they are manufactured is about the same. It is certainly wrong for a large business, manufacturing a large variety of products.

The main drawbacks of this type of full costing in a large company are twofold:

1. First, the total cost computed for the products are probably wrong, as it will appear in the following pages. The comment of Gokrat mentioned earlier about

---

[2] The manual (Ref. [19]) mentions that overheads comprise about two-thirds of the total in-plant cost of most defense contractors. This was in 1975 and the percentage increases continuously.

the deterioration of the US productivity is probably caused by this factor, simply because decisions are based on costs: if the cost figures are wrong, decisions also are.
2. Second, this procedure demotivates the product manager. Whatever the effort he/she deploys for reducing his/her costs, he/she is "submerged" by a large cost on which he/she has no control at all …

Let us see how the cost analyst who receives a proposal from a company can guess if this company sets prices to its products according to this simple rule. The proposal mentions the costs of material (multiplied by the material overhead rate), plus the added value by the company; if this added value is computed from direct hours for instance, the proposal mentions this number of hours and multiplies it by a single rate which take into account the direct cost and the manufacturing overhead. Such a rate is called "undifferentiated" which means that every hour is rated the same way. In such a case, you may be pretty sure that the "rough" solution was used; the number of hours may be correct, but the way they are converted in € or in £ or in $ may be challenged.

## 2.4.2 The "Standard" Solution

We described here, with a lot of simplifications, the standard solution which was implemented for solving the problem in many large or small companies.

The basic idea of this standard solution (it is called "standard" because it is largely used) is that in a large company a lot of different services are provided and these services are very inhomogeneous and cannot therefore be allocated to products according to a simple rule. Therefore one solution could be to decompose the company into several "sub-companies", the criteria for this decomposition being that the services rendered by each one should be homogeneous enough to allocate its cost in a satisfactory way.

These "sub-companies" are called "cost centers", or "profit centers": choose the name you prefer. As a cost analyst, we prefer the first one. The number of the cost centers, which is an important decision to make, might be small or large to very large (thousand cost centers inside the same company is not exceptional, for instance in the automobile industry); let us call it $M$.

In order for the system to work properly, it must be possible to "isolate" (from a cost accounting point of view) each cost center. This means that its inputs and outputs have to be measured. There was a lot of discussion about the experts on how these transfers, once physically observed, should be quoted; our subject being only to introduce this cost accounting system, let us remain as simple as possible.

The cost center $i$ receives two kinds of resources:

1. Resources directly bought for it from the outside world (salary, energy, transportation, etc.). Let us call them its "apparent" cost $R_i$ (the word "direct" could have been used, but it is avoided here in order not to confuse it with the direct costs of the products). They are the resources it buys from the outside world.
2. Services it receives from other cost centers (procurement, machine hours, maintenance, etc.). Let us call the cost of the services received from cost center $j$, $S_{i,j}$.

The same cost center $i$ delivers two kinds of services:

1. To the projects or products when it works for them; let us call the cost of the services received by project $p$, $S_{p,i}$.

2. To the other cost centers; let us call the cost provided by cost center $i$ to cost center $j$, $S_{j,i}$. Of course, in general $S_{i,j} \neq S_{j,i}$.

From the point of view of **the manager of cost center $i$**, its total cost is obviously equal to:

$$T_i = R_i + \sum_j S_{i,j} = \sum_p C_{p,i} + \sum_j S_{j,i}$$

From the point of view **of the manager of product $p$**, its full cost is given by:

$$F_p = D_p + \sum_i C_{p,i}$$

where $D_p$ is its direct cost.

From the point of view of **the plant manager**, his/her capacity cost $K$ (with no index) is given by:

$$K = \sum_i R_i$$

How are the $S_{p,i}$ and $S_{j,i}$ (the cost of the services given to projects and other cost centers by cost center $i$) determined? Their values depend, of course, on the activity of the center. The computation starts with the definition of the cost center potential: this potential is the amount of services, generally measured by the number of hours (labor, machine, etc.) it can deliver during a given accounting period. The $S_{p,i}$ and $S_{j,i}$ are computed from the knowledge of the potential percentage used for each project or delivered to each cost center.

Immediately a problem arises: what happens if the potential of services during an accounting period is not fully consumed? A cost center has a decided (by management) potential to produce these services; let us call it $k_i$: for instance $k_i = 25\,000$ h of something (labor, machine time, etc., depending on its activity). If the potential is not fully consumed two solutions are possible:

1. The $S_{p,i}$ and $S_{j,i}$ are adjusted (multiplied by a factor) in order to make the potential appears fully consumed.
2. A special "project" is created for recording the unused potential.

The first solution is often selected, probably because managers of cost centers do not like to make the unused potential to clearly appear. But, independently on the fact it is always interesting for the plant management to know it, it makes the cost of the products fluctuate from one accounting period to the other one, as a function of the activity.

Let us return to the computations: let us call $\alpha_{j,i}$ the proportion of the resources of cost center $j$ it receives and $\alpha_{p,i}$ the amount of resources it delivers to project $p$; if for instance cost center $j$ has a potential of 3000 working hours and among these hours 300 were employed to provide services to cost center $i$, then $\alpha_{i,j} = 10\%$. The full cost of this cost center $i$ is then computed as:

$$T_i = R_i + \sum_{j \neq i} \alpha_{j,i} T_j$$
$$T_i = \sum_p \alpha_{p,i} T_i + \sum_{j \neq i} \alpha_{i,j} T_i$$

which delivers $M$ linear equations:

$$R_i + \sum_{j \neq i} \alpha_{j,i} T_j = \sum_p \alpha_{p,i} T_i + \sum_{j \neq i} \alpha_{i,j} T_i$$

where only the $M$, $T_i$ are unknown and $\sum_p \alpha_{p,i} + \sum_j \alpha_{j,i} = 1$. This system of equations can easily be solved, delivering the $T_j$.

This way of computing full costs are interesting because it delivers three kinds of information:

1. *The total cost of the products:* $F_p = D_p + \sum_i C_{p,i}$. This was the first motivation of this accounting system.
2. *The total cost of each cost center:* $T_i = R_i + \sum_{j \neq i} \alpha_{j,i} T_j$. Consequently it allows easily the delegation process: each cost center manager is considered a manager of a small plant and his/her performance can be measured.
3. *The rate of each cost center:* $\rho_i = T_i/k_i$, for instance 145€/h. This allows for comparing the performance of one cost center to the other ones, internal or external to the company; it may be an important information for subcontracting decisions.

Practically the $M$ linear equations are not directly solved. When this accounting process was created no computer existed which would made possible to solved let us say 200 linear equations and it must be recognized it would have been an impossible task to solve it manually. The solution which was found can be called the "waterfall solution": the cost centers were organized as a hierarchy. At the top of the hierarchy were cost centers which did not (or could not) receive any services from other cost centers; at the second level were cost centers which could receive services from the first level but not from the other level; etc. The hierarchy could have as many levels as management decides, the principle being that a cost level at level $l$ could receive services from cost centers at levels $l - p$ (whatever $p$) but could only give services to cost centers at levels $l + q$ (whatever $q$). This solution was manually feasible; nowadays the normal procedure of simultaneously solving the set of linear equations can be used and has already been used.

## 2.5 The Method ABC

The method ABC (the "activity-based costing") can be considered as a refinement of the standard cost accounting system.

The idea is that workers which belongs to the same cost center, and therefore provide "labor hours" to projects or other cost centers, may have to carry out different activities and the cost of these activities should be properly appointed to the projects, as the following example illustrates.

### An Example

Suppose a plant produces two different products, let us call them A and B. A is sold for €50 per unit and B for €40 per unit. We suppose that only one cost center is involved in the manufacturing process and that the overhead of this cost center is €2750.

The plant manager would like to start the production of a new product; as the potential of the plant is limited, this cannot be done without deleting one of presently manufactured products. Of course, the plant manager wants to delete the less profitable product. He consequently asks to the accountant the profitability of each product.

## A First Computation

The accountant has a good record of the production cost. From his records, he computes that for the production of two batches of 10A and 100B which occurred during the last accounting period:

- The direct costs are:
  For product A: material (including material overhead): €10;
  direct labor: 5 h @ €10/h or €50.
  For product B: material (including material overhead): €100;
  direct labor: 50 h @ €10/h or €500.
- The indirect cost amounts to €2750.

For this cost center, the overhead rate is computed on the direct hours. Consequently the accountant computes a rate of 2750/55 = €50/h.

Consequently the accountant computes:

- For product A a full cost for the batch of €3100 (direct cost €600 + indirect cost 5 × 50 = €250) and therefore a unit full cost of €31 and a margin of €19 per product sold.
- For product B a full cost for the batch of €310 (direct cost €60 + indirect cost 50 × 50 = €2500) and therefore a unit full cost of also €31 and a margin of €9 per product sold.

The plant manager should therefore delete product B.

## A Second Computation

Before sending this information to the plant manager, the cost accountant checks the direct costs and realizes it is correct. Doing the same investigation for the indirect cost, he questions the way the rate is computed and decides to get a better understanding of the validity of the procedure. Following a discussion with the cost center manager, he realizes that the overhead cost should not be distributed equally on all the direct hours because it can be split into two components:

1. *Preparation of each batch* (changing the tools, tuning the machines, etc.) before the manufacturing really starts.
2. *Maintenance of the machine* during the manufacturing process.

A quick investigation reveals that the preparation of the production line before each batch costs €270 and the machines maintenance €1670. As four preparations were performed during the last accounting period, the overhead cost can be analyzed the following way:

$$2750 = 270 \times 4 + €1670$$

Consequently he decides to distribute the indirect cost according to the two activities performed by the cost center:

1. €270 per batch;
2. €30.36 per labor hour (30.36 × 55 = €1670).

Product A requires one production batch only, whereas product B requires three. The computation of the full cost of both products is now given by:

| Product A | Direct labor | €60.0 |
| --- | --- | --- |
| | Preparation of one batch | €270.0 |
| | Overhead | €151.8 (€30.36 × 5) |
| | Total | €481.8 |
| Product B | Direct labor | €600.0 |
| | Preparation of three batches | €810.0 |
| | Overhead | €1518.2 (€30.36 × 50) |
| | Total | €2928.2 |

Product A has now a unit production cost of €48.18 and each product sold creates a margin of €1.82, whereas product B has unit production cost of €29.28 and therefore each product sold creates a margin of €10.72.

According to this computation, product B is the most profitable and product A should be deleted.

Of course this illustration is oversimplified, as the quantity sold should be taken into account, as well as the forecasted sales! Nevertheless it shows that the accounting system by itself may produce bad decisions.

### Conclusion

This example illustrated the fact that in some situations, when the work produced by a cost center cannot be considered as homogeneous enough, indirect cost should not be uniformly distributed.

This was the origin of the ABC costing method. The logic of this method is that "products are manufactured by activities, and activities consume resources".

Going down to the activity level may represent a large investment for a company; it may be necessary, depending on the breakdown of the company in cost centers, for all of them, for a few of them or for none of them. But the problem deserves examination. The subject has been largely presented in several books among which one can mention Ref. [33].

## 2.6 What Can the Cost Analyst Expect?

The purpose of this chapter is to introduce the cost analysts to the way costs are "measured". We hope we made it clear that cost accounting is not as so simple as it may appear at the first look and required the help of professional people.

Beyond this discussion the cost analyst may expect an answer to the question he/she may ask: "How accurate are the costs information I receive?". Cost information received from the cost accountant is precise to the cent. Does this mean that this information is perfectly reliable?

There is not easy answer to this question, for several reasons.

The first reason comes to the fact that, in the cost domain, we are generally, as cost analysts, not so much interested in absolute precision as we are in relative precision: making a measurement error of €10 on a product which is sold for €20 really means we know nothing about the product cost. Making the same error on a product sold €1000 means a high precision, whereas on €10 000 is not credible at all; if it were, it means that the cost accounting system is over-designed!

The second reason is the fact that it depends, as it was discussed throughout this chapter, on the decided cost accounting system. The reader may have perceived the fact we have little faith in the result of a "rough" full costing and that the credibility of the "standard" full costing depends first on the variety of the products and services sold by the company, second on the breakdown of the company in cost centers (Are the decided ones homogeneous enough?).

The third reason is that we do not deal with physical measurement and a high precision on each task is not realistic: never forget that an accuracy of ±5% means a precision of ±3 min on 1 h: not so many people are able to attain such a precision!

The fourth reason is human: as one person said to us once: "we certainly have the best accounting system of the continent. But there are so many ways to cheat it that we certainly do not know our costs with a better accuracy than the other companies".

The fifth reason is purely mathematical: suppose a project is made out of 100 tasks and, in order to simplify the discussion, that each task is worth €100 and is known with a precision (a standard error) of €30 (which is rather poor) and that all tasks are, from a cost point of view, independent. The total cost of the project is equal to €10 000. The standard error of the total cost, assuming the tasks are independent is given by[3]:

$$(30^2 \times 100)^{0.5} = 300$$

and the relative precision of the total cost is then equal to 3%, which is quite good! It means we do not have to know the cost of each task with a high precision.

The sixth reason is the fact that "measured" costs are most often "allocated" costs: when a job has to be done, a time is allocated to it. Most often people in charge of the job do not record the exact time it took to really make the job, but the time which was allocated to do it, even it the real time was shorter (on the contrary if the real time was longer than allocated, then the real time is generally recorded). It means that the accuracy of the time is difficult to appreciate.

Taking into account of these remarks explains that the accuracy with which the cost of the products is known is not extremely high and that a 5% figure is realistic. It may be less for similar products made by the same company, worse for different products made by different companies. In this last case, the differences in the accounting system (the way costs are distributed to the products, the depreciation times, the nature of costs considered as direct, etc.) explain why the costs measured by two different companies for exactly the same product can differ by more than 5%.

There are two consequences of these remarks:

1. Do not expect a cost model to give estimates with an accuracy better than about 5%, at least for the manufacturing costs (the accuracy is even less for engineering costs).

---

[3] For independent values the standard deviation of a sum is equal to the square root of the square of the individual standard deviations.

2. A difference of less than 5% between measured and estimated costs is not really significant.

Of course the results are even worse if you compare prices between two companies even for the same products because the pricing policy (going from cost to price) is not known.

## An Objection Often Heard

We often heard the following comment: "if you cannot estimate costs with an accuracy better than 5%, your estimates cannot be used because we have to commit ourselves to a fix cost and that, on this fix cost, our profit margin is 2%"!

Everybody is of course right. So, how this contradiction can be solved?

A preliminary remark is that the cost which is estimated is the "nominal" cost, based on past experience. If the level of uncertainties for this project is not negligible, it has to be taken into account; we will see in Chapter 18 how it can be done; furthermore going from cost to price requires an analysis of the risks and contingencies: if this amount is higher than what was observed in the past, it has to be taken into account.

The second remark is more fundamental and we will illustrate it on an example. Suppose that we are interested in the time needed to go walking to the train station. We made in the past several experiences which revealed that this time is well represented by a Normal distribution with an average time of 1 h and a standard deviation of 3 min (which depends on the weather, the traffic, the luck at the traffic lights, etc.). The estimate we give is $1\,h \pm 3\,min$.

Now we are said there is a train in 58 min and that we have to reach it! In order to meet this objective we prepare a plan (at that time we should be at this place, etc.). When the plan is made, we depart and enter in a "control" phase: every 5 min we compare the place where we should be to the place where we are and react accordingly (from stopping for a coffee cup if it is possible, to running if it appears we are late) in order to meet the objective.

The lesson of this short story is that there is a fundamental difference between preparing an estimate and aiming at an objective. The estimator does his/her best taking into account the results of several past experiences, whereas the project manager organizes his/her project in order to meet an objective: it is quite possible to meet an objective with an accuracy of 0.1%, even if the estimate was at 5%. Both things are completely different. Estimates are made, as previously mentioned, to prepare decisions but are not substitutes to the management process.

# 3 Overview of Cost Estimating

## Summary

This chapter starts with a general presentation of the various approaches used for cost estimating.

It briefly describes the usual methods to get them, insisting on the fact that any cost estimate is based on comparisons, or analogies. Analogies can be quantitative or qualitative, the trend, with the disposal of computers, being to shift more and more often to the quantitative approach, which is the general meaning of parametric cost estimating.

However, it also shows that this technique is not synonymous of formulae: it has being used for quite a long time with tables or graphs, formulae being only the last avatar.

The most common ways for cost estimating are – beyond the mere catalogs – the heuristics, the expert judgment and, when the volume of the database is large, the case base reasoning (CBR) which is based on data organization.

Modelization is the current technique to operate parametric cost estimating.

A brief history is presented, only for showing that it has been, for centuries, in the mind of everybody who was concerned by cost estimating.

The foundations and the present definition of the technique (at the product or equipment level) are then introduced.

Eventually a definition is given of the two major ways of implementing this technique: specific models and general models.

In order to prepare an estimate you need (manufacturing a product is used for illustrating the concepts, but other examples could as well be taken: software development, construction, etc.):

1. To describe the product; the product description has two sides:
   - The **nature** of the product. It is the response to the question: "What is it?" Is it a bicycle? a lawn mower? a satellite? a spare part? The nature of the product includes the quality level.
   - The product **size**. This is a very important piece of information which has always to be considered: if only one variable is used (assuming the nature of the products to be considered is the same), it has to be the size.
2. To define the phase to be estimated: are you going to estimate the cost of the preliminary studies? of the full development? of the production? etc.
3. To describe the environment in which the phase will take place. This environment will have to take into account the tools you have, plus the constraints.

In this chapter we consider only the product description.

As cost estimating is based on a comparison between the result of past experiences and what has to be estimated, the concept of distance between products is a very important one. As the product size can be taken into account independently, distance is principally concerned by the product nature. A central problem for the cost estimator is to appreciate this "distance" between two products; most often the appreciation in not conscious, but cost estimators are encouraged to investigate the subject.

This subject is not so obvious as it seems. Take the example of an electronic box (such as a receiver). You may consider that the distance between two boxes is small if the receivers have about the same sensitivity, about the same bandwidth and work in a not too large frequency band. However, a very important cost driver is forgotten: it is the nature of the electronic components! Can we really compare two electronic boxes, one with components at $1\,\mu$ and another one with components at $0.2\,\mu$? Certainly not and this is more important than any other cost driver. For electronic boxes, the nature of the components has to be taken into account (or at a minimum, the year during which the box was developed).

The same comment can be done for mechanical items: the material to be used is an important cost driver.

## 3.1 Product Description and Approach

This is the first thing to do: the object, whatever it is (hardware, software, building, tunnel, etc.) has to be described. This description has two faces:

1. What do we look at?
2. How detailed is the description?

### What Do We Look at?

There are three ways to look at a product. We can look at:

- What it does? This is the "functional" approach: a product is made to perform such or such function, at a certain level, in a specified environment. This is sometimes called the "customer's approach", because, obviously, it is what the customer is interested in. This approach is characterized by three sets of "parameters":
  - the list of the functions it has to perform,
  - the level of each function,
  - the environment in which the "product" has to perform the functions it is built for. This can be referred to as the "quality level" which is expected.
- What is it? This is the descriptive approach, sometimes called the "engineer's approach", because it is the way engineers "think" about a product. It describes how the engineers design the product for fulfilling the functions already mentioned. This approach is characterized by:
  - the drawings, which include the product shape and the tolerances, the list of parts, etc.,
  - the material(s),
  - the controls they consider that have to be done to be sure the product will perform as expected.
- How is it made? This is the "constructive" approach, sometimes called the "manufacturer's approach". The purpose is not any more to produce a function but to

produce what the engineers have designed and specified. This approach is characterized by:
- the list of material(s) to procure,
- the activities to perform for both the production and the controls. This approach, because manufacturing is achieved by activities, is oriented to activities: a visit in a plant clearly shows that.

These three approaches used to be sequential: the functional approach was, should have been, the first to appear. Then the design starts, followed by manufacturing. Nevertheless there is a trend, as can be seen in the cost as an independent variable (CAIV),[1] to consider them simultaneously.

The cost estimator may also work – whatever the approach – at the top level (this is called the "global approach"), or go down, maybe at the lowest level, called the "detailed approach".

Generally the word "detailed" is used for describing a "method" to estimate. We do not think it is correct: it is quite possible to estimate at a very detailed level, following a functional approach, or a descriptive approach or a constructive approach. The word "method" is used here to describe how one can go from a description, whatever it is, to a cost: method is not synonymous to level of detail.

When we go down from the global to the detail, we are not forced to stay in the same approach: the arrow on Figure 3.1 reminds that. Starting at the top-level functional approach, it is possible to shift to the descriptive approach and to stay some time in this approach. If the information is available for some parts, it is also quite possible to investigate for these parts the constructive approach.

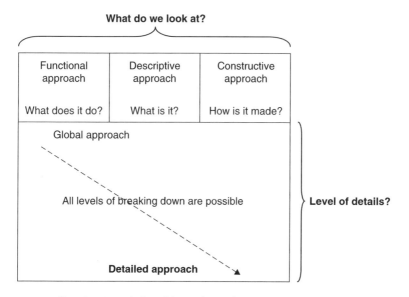

**Figure 3.1** Description of the product and cost estimating approach.

---

[1] CAIV stands for "Cost As an Independent Variable." It is a procedure for reducing the engineering phase of a product and its duration, for reducing its costs of production and maintenance, and, more generally to provide the functions at the lowest cost.

Consequently the framework we propose is not a constraint: it is just a description of what happens. The only constraint is that the process goes in one direction only: this why the arrow on Figure 3.1 goes from left to right: if you abandon, for some function, the functional approach, there is very little chance that you will return to it.

The only thing we insist on is that breaking down the product should not be called a method. The vocabulary in itself is not the major issue, but we have to clearly explain the way we work: if we are not vigilant to the vocabulary, we add to the confusion.

## 3.2 The Usual Methods

The classification we propose here is different from what is generally presented. The reason comes from the fact many cost estimating practices use the same techniques, but apply them to different things; this generates a lot of confusion, especially among engineers. As we believe that engineers are – or should be – the focal point for cost information (after all they are the ones who generate the cost!), we think that the classification of cost estimating practices should be based on the techniques only.

Cost estimating practices seem to vary a lot from one company to another one. However, a typology of cost estimating methods reveals three major different approaches, even if approaches cover a lot of variations, ...

Textbooks distinguish between analogy, analytic, parametric cost estimating techniques, plus sometimes many other words. So many distinctions bring confusion, ... Let us try to clarify the subject.

First of all it is clear that the only way we can predict the future is by comparison. This comparison may be extremely simple, and even not necessary, if the thing we have to estimate has already been made: in such a case, we just have to look to a catalog (which is the old version of a database) to find the cost – more often the price – of the product we are interested in. If it is not the case, there are two main ways to make an estimate:

1. If we want to stay at the product level, we will try to compare the product with something which has already been done before. This is generally called "**analogy**":
   – Analogy may take very different forms from expert opinion to establishing quantitative relationships.
   – Analogy is not always very easy; we will see that the problem becomes worse and worse if the product(s) available for such a comparison is less and less "close" to the product we want to estimate. We will return to that in the next chapter.
2. If we do not want to – or cannot – stay at the product level, the only thing we can do[2] is to analyze how the product will be made; this is generally called for obvious reasons "**analytic**" (other words are "grass roots", ...). So we shift now from the "what" to the "how". Looking to the "how" requires to make the list of all the activities which will have to be carried out to make the product (this is the "operations sheet" used in the manufacturing industries). Generally speaking each activity implies:
   – Some material (from raw material to elementary parts such as nails or any other thing). The cost of these items, bought from external sources, is found in catalogs.
   – Some equipment and labor, for which we have – this is the purpose of the cost accounting – the hourly rate. That is fine, but our problem now is to estimate

---

[2] Unless we can break down the product into "sub-products" for which a comparison is possible. This is obvious.

the time we need them. How do we do that? Unless we find this time in a catalog, we have to return to analogy.

If the basic information we need cannot be found in a catalog, all cost estimating is based, at one level or at another on analogy.

Analogy is therefore a central process in any forecast, about cost or any other thing. Analogy can be of the following:

- *Qualitative*: This is the normal process you use everyday when you want to buy something new to you, from a new meal to a new house or a new equipment for which you do not have the catalog. The logic is very simple: from your experience of products, maybe similar, maybe different, you try to answer the question "how much is that going to cost?". In the industry it is referred to as "expert opinion". It can be very accurate if the thing to be estimated is right in the domain of the expert; if it is not, methods have been developed (such as the Delphi method) in order to reduce the subjectivity of the expert in front of something rather new to him. It is also the "method" you use, as a manager, when you receive an estimate or a quote: "is it reasonable?".
- *Quantitative*: Qualitative analogy can be sufficient if two conditions are met:
  - The number of "cost drivers" (we will use a lot this term: a cost driver is just something that may change the cost; if you are looking to a next house, you know that the cost drivers are the size, the location, the quality of the construction, etc.) is very limited. The human mind is not very good in taking into account, at the same level, several factors that can influence what it is thinking about (this is well known in behavioral sciences as "limited rationality"). This helps explain why the size has such an importance in the cost estimating community (there are other reasons which are discussed below).
  - The relationship between the cost drivers and the cost is about linear. The human mind is not good at all at manipulating, without any support, non-linear relationships. It tries as soon as it can to play with linear ones.

Unfortunately, it is rather rare that both conditions are met.[3] In such a case, we turn to quantitative analogy.

By "analogy" we generally mean comparing the product to be estimated to another product. But the comparison will obviously more fruitful if we can compare it, *simultaneously*, to several products: all these products will then contribute to the cost estimating process. But, in order to palliate our "limited rationality", we are forced to develop methods allowing this simultaneous comparison; it is here that the mathematics enter the picture. Mathematics must be seen as just the way we found for compensating the limited capacity of our brain to make complex comparisons.

The mathematics we are going to develop and use for cost estimating purposes are called: "parametric cost estimating".

In this book quantitative analogy will be referred to as "parametric cost estimating".

---

[3] The fact that many "cost estimating relationships" or CERs, published in the literature contain only one cost driver is not a proof of anything of that sort.

**Table 3.1** MTM data table for "Reach".

| Distance moved inches | Time TMU | | | | Hand in motion | | Case and description |
|---|---|---|---|---|---|---|---|
| | A | B | C or D | E | A | B' | |
| 3/4 or less | 2.0 | 2.0 | 2.0 | 2.0 | 1.6 | 1.6 | A Reach to object in fixed location |
| 1 | 2.5 | 2.5 | 3.6 | 2.4 | 2.3 | 2.3 | or to object in other hand or on |
| 2 | 4.0 | 4.0 | 5.9 | 3.8 | 3.5 | 2.7 | which other hand rests. |
| 3 | 5.3 | 5.3 | 7.3 | 5.3 | 4.5 | 3.6 | B Reach to single object in location |
| 4 | 6.1 | 6.4 | 8.4 | 6.8 | 4.9 | 4.3 | which may vary slightly from |
| 5 | 6.5 | 7.8 | 9.4 | 7.4 | 5.3 | 5.0 | cycle to cycle. |
| 6 | 7.0 | 8.6 | 10.1 | 8.0 | 5.7 | 5.7 | |
| 7 | 7.4 | 9.3 | 10.8 | 8.7 | 6.1 | 6.5 | C Reach to object jumbled with |
| 8 | 7.9 | 10.1 | 11.5 | 9.3 | 6.5 | 7.2 | other objects in a group so that |
| 9 | 8.3 | 10.8 | 12.2 | 9.9 | 6.9 | 7.9 | search and select occur. |
| 10 | 8.7 | 11.5 | 12.9 | 10.5 | 7.3 | 8.6 | |
| 12 | 9.6 | 12.9 | 14.2 | 11.8 | 8.1 | 10.1 | |
| 14 | 10.5 | 14.4 | 15.6 | 13.0 | 8.9 | 11.5 | D Reach to a very small object or |
| 16 | 11.4 | 15.8 | 17.0 | 14.2 | 9.7 | 12.9 | where accurate grasp is required. |
| 18 | 12.3 | 17.2 | 18.4 | 15.5 | 10.5 | 14.4 | |
| 20 | 13.1 | 18.6 | 19.8 | 16.7 | 11.3 | 15.8 | |
| 22 | 14.0 | 20.1 | 21.2 | 18.0 | 12.1 | 17.3 | E Reach to indefinite location to get |
| 24 | 14.9 | 21.5 | 22.5 | 19.2 | 12.9 | 18.8 | hand in position for body balance |
| 26 | 15.8 | 22.9 | 23.9 | 20.4 | 13.7 | 20.2 | or next motion or out of way. |
| 28 | 16.7 | 24.4 | 25.3 | 21.7 | 14.5 | 21.7 | |
| 30 | 17.5 | 25.8 | 26.7 | 22.9 | 15.3 | 23.2 | |
| Additional | 0.4 | 0.7 | 0.7 | 0.6 | | | TMU per inch over 30 in. |

This classification does not mean that a tool belongs exclusively to one category. Some tools use concepts belonging to several. But it is convenient, from an academic point of view and for helping the reader to understand the concepts, to distinguish different categories.

Distinguishing between analogy, analytic, etc. does not really matter. Cost estimating, either at the product level or at the activity level, can only be done "parametrically", unless the product or the activity is already present in a catalog.

Parametrics is the central concept in cost estimating.

Parametric cost estimating has been there for many years. However, its practice would gain at being formalized. It is a powerful concept and extremely efficient if properly used.

## Parametric Cost Estimating Does Not Mean Only Equations, ...

Before every cost estimator could use a computer, parametric cost estimating used just tables, as illustrated in Table 3.1 published in the *Method Time Measurement* (MTM) handbook.[4]

---

[4] *Motion and Time Study. Design and Measurement of Work.* Ralph M. Barnes [7] is a good introduction about this use.

This table clearly illustrates what we mean by parametric cost estimating: it includes two parameters:

1. *A quantitative parameter*: the distance where the object to be reached is located.
2. *A qualitative parameter (designated by the letters A, B, C, D and E)*: how definite is the location?

and directly gives the cost (here the time in TMU, or time measurement unit of $10^{-5}$ h or 36 ms). It also shows, as previously mentioned that parametric cost estimating is not dedicated to the global approach: it is difficult to get a more detailed approach!

Another presentation largely used by our parents was the graph, illustrated below: the following example (from Ref. [3]) shows how it is possible to compute the cost – here the time given in hundredths of minutes or "cmn" – to bore four holes of the same diameter in gray iron. The effort is divided into two activities:

1. Positioning the material on the boring machine (Figure 3.2). On the example we can see that if the total distance between the holes is 120 cm and the number of holes is equal to 4, then the allocated time is 14.3 cmn or 8.6 s.
2. Boring the holes. Figure 3.3 is the graph which is used for one hole. From left to right we find:
   – the hole diameter (11.5 mm in the example);
   – once again the diameter, on a special rule for material thickness larger than 10 mm;
   – this gives the allocated time for a standard depth of 10 mm;
   – then we take the real depth of 50 mm;
   – which corresponds, for the soft steel, to 154 cmn;
   – from this point, a horizontal line gives the allocated time for different materials,[5] 110 cmn for gray iron.

The total allocated time for boring the four holes is then 454.3 cmn or about 4½ min.

This example is also typical of parametric cost estimating: it uses several quantitative parameters (distance between the holes, number of holes, diameter of the holes and their depth) and a qualitative one (type of material to be bored).

Nowadays, with the generalization of the computers, parametrics is one of the process by which (continuous) cost models are built.

The other way is "curve fitting". But this technique can, practically, only be used when just one parameter is involved. It consists in two steps:

1. Plotting the data on a graph.
2. Finding the curve – or the equation of this curve, this is exactly the same – that represents the data; "representing" the data means that no so much information is lost when the data are replaced by the curve or the equation. This is generally done manually (for the curve) or by trials (for the equation).

Quite obviously this process can only be done on a piece of paper or on the screen of a computer: only one parameter at a time can be used.

Curve fitting is still used, even by parametric models builders when they get enough data to work on or when they have a good idea of what the behavior of a variable should be, and when statistical packages do not give the result they are looking for. There is nothing wrong about it.

---

[5] The eight vertical lines allow you to compare the machinability of different materials.

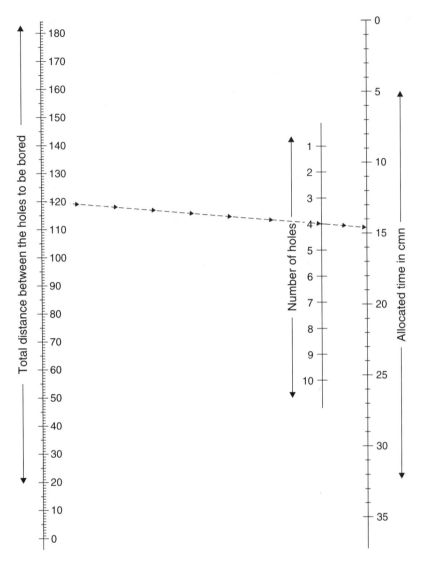

**Figure 3.2** Standard time for positioning the material to be bored.

Let us return to the methods of cost estimating. We reserve this word "method" to the process which goes from the description, whatever the approach, whatever the level of detail, to the cost.

The Ariane thread is the distance between past experiences and the new one.

### 3.2.1 Heuristics

Heuristics are mainly used by managers in order to get a rough order of magnitude of a project cost. They are simple rules which may produce a realistic cost or not,

# Overview of Cost Estimating

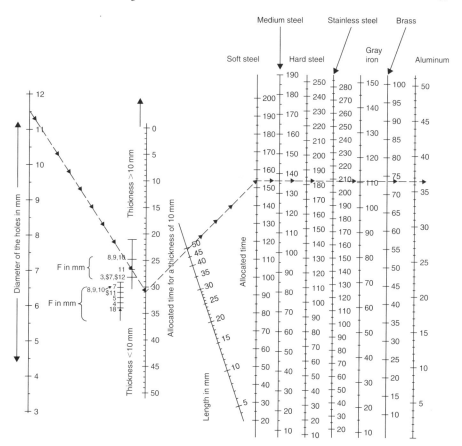

**Figure 3.3** Standard time for boring one hole.

the problem being that we are never sure about the quality of the result ... But they can give a good order of magnitude of the cost.

As everybody knows that cost is a function of the size of the thing to be estimated, heuristics generally present a linear relationship between the cost and the product size; more will be said in the next chapter about it. At this stage, one may observe that the quantification of the size depends on course of the product type: therefore heuristics are often expressed as cost per kilogram (for hardware), cost per square meter (for buildings), cost per cubic meter (for tunnels), ... In software one generally used the inverse relationship between cost and size: it is called "productivity" and is given by the number of instructions (or DSI, which stands for "Deliverable Source Instruction") per man × day.

Let us consider for a moment results given in the software industry (*Source*: NASA, published by ISPA):

Program: Shuttle
   in board software:      1.3 to 2.1 DSI/man × day
   ground software:       5.1 to 22.6 DSI/man × day

Program: Gamma Ray Observatory
    in board software:    1.4 to 3.3 DSI/man × day
    ground software:    7.0 to 15.3 DSI/man × day
Program: Spacelab
    in board software:    2.8 to 3.4 DSI/man × day
    ground software:    9.6 to 12.6 DSI/man × day
Program: Galileo
    ground software:    5.2 to 27.0 DSI/man × day

What can be immediately noticed is a clear distinction between two families of software: in board software (productivity goes from 1.3 to 3.4 DSI/man × day) and ground software (productivity goes from 5.1 to 27.1 DSI/man × day). The difference may come from the quality level and the amount of testing. It is also interesting to note that the productivities are very scattered; this scattering may come from at least two factors:

1. The fact that they refer to very different software. This shows that such ratios, and we will insist many times on this subject, can only be used inside very homogeneous "product families".
2. The fact that the size of the software may be very different. Using a ratio assumes that the cost grows linearly with the size, which is certainly not the case.

### 3.2.2 Expert Judgment

The expert knows that several factors, let us call them $f_i$, influence the cost. In order to go from the factors to the cost, he/she has to give them a weight, let us call them $p_i$, which he/she has to explicit: what is the absolute effect of this factor? What are the relative effects of these factors?

The problem is, when the number of factors is larger than 2, to make sure that the expert is consistent. Suppose this person thinks that factor $f_i$ is four times more important than factor $f_j$, but that factor $f_k$ is three times less important than factor $f_j$. Answering to another question, he/she thinks that factor $f_i$ is two times more important than factor $f_k$. Is this scale consistent?

It is not too difficult to answer this question when the number of factors is small, for instance limited to 3. It becomes much more complex if this number becomes large. With seven factors for instance you are facing 21 relative influence of couples $(f_i, f_j)$ and the problem of their consistencies is more difficult to solve. Methods have been developed for doing that. More will be said about them in Volume 2.

It is also possible to use several experts. The problem is then to reconciliate their judgments.

### 3.2.3 Data Organization and CBR

The idea is, once you have collected a lot of data, is to organize them. Data organization follows three objectives:

1. Searching for a product close enough to the product you have to estimate. This is the first and simplest way to prepare an estimate. If the database is large, several

techniques can be used, the CBR, which stands for "case base reasoning", being an interesting example.
2. Grouping the data into product families. This step is necessary if the product you have to estimate is similar to several products of your database, but not similar enough to be able to directly select one of them for a close and direct comparison. This will be the subject of Volume 2.
3. Searching in the database a few products for adjusting a general model. This question is the exact opposite of the previous ones: you are looking for a set of products which are dissimilar enough to explore various aspects of a general model.

## 3.3 Modelization (Parametric Cost Estimating)

### 3.3.1 Definitions

*A Preliminary Definition*

Suppose we have to make a cost estimate of "something", whatever it is: it can be a product, a building, a road, a tunnel, an activity or anything else. The previous sections were a reminder of the usual ways to proceed to get it.

Looked from a distance, parametric cost estimating is nothing else than the formalization of the heuristics or the expert opinions. Both start from a description – or even an idea – of what is to be estimated, and by looking at its size, its "complexity", its manufacturing environment, etc., try, by using experience generally informal, to get an immediate estimate, which can be just an order of magnitude or, if a similar thing was done not so long ago, a figure rather accurate.

The logic is then obvious and can be expressed the following way: heuristics and/or expert opinion may provide some realistic figures, but two important points are missing:

1. There is no traceability and therefore it seems difficult to improve the process.
2. It is impossible to transfer the knowledge to somebody else: when the expert is not available, nobody can be substituted to him/her. When he/she leaves the company, there is a large amount of knowledge which disappears.

Can we try to formalize this process in order to solve these difficulties?

One solution can be found in what is called the "expert system"; its purpose is, through interviews, case studies, etc., to "extract" the knowledge of the expert's mind and to formalize the way he/she thinks. Expert systems are built as sets of rules around an inference "engine" which is able to link the rules together. This solution, which gave interesting results in the domain of diagnostics, can be used if the production process is very stable. It involves a very important investment.

The other solution consists in directly "modelizing" the relationship the expert establishes between the cost drivers and the cost:

- The expert starts from facts: as it is the basis of everything we better collect data. Once this is done, we will also start from these facts.
- The expert, even he/she has difficulties to explain in a detailed way how he/she goes from the facts to the cost, extracts from these facts a representation of the

cost behavior. We will do the same thing; this is of course the most delicate part of the process, which we will call the "model building" phase.
- He/she then exploits this representation and the description of the object to be estimated in order to get an estimate. This will be called "using the model".

At this stage, it is then possible to get a very **general definition** of parametric cost estimating as the process of formalizing the way the expert thinks in order to make an estimate. This formalization presently uses the construction of a model. The presentation of this model may be anything: it can be a table of values, or a chart, or a mathematical relationship.

### The Present Definition of Parametric Cost Estimating

First of all the words "parametric cost estimating" refer nowadays mainly to products, the term "product" being used in a very extensive meaning, from parts to equipments and systems, buildings, etc. Estimators who work at the activity level do not use the term, although they could according to the general definition.

Second tables and charts are not anymore in use, due to the large diffusion of computers.

The **present definition** can then now be written as:

*Parametric cost estimating is the scientific approach for going directly, by using mathematical relationships, from the product description to its costs.*

A few comments about this definition:

- The word "scientific" has been added to mean that the process is now well documented and formalized (which does not mean at all it is definitively finished!).
- The word "directly" means something very important about this technique: the usual process of making the list of activities which have to be carried out and estimating the cost of each is bypassed. The purpose is really to go directly from the product description to the cost, taking – when it is necessary – the production environment.
- Mathematical relationships is now the standard way to implement the technique.
- "Costs" has to be plural, as the cost of the various phases through which a product could go during its life can be parametrically estimated: development, preproduction, production, delivery, training, maintenance, decommissioning, etc.

### What Is a Parameter?

A parameter is a descriptor of the product or its development/production environment which influence the cost.

Mathematically, we will call it a variable.

The word will be redefined and deeply investigated in Volume 2.

### 3.3.2 A Brief History

Parametrics cost estimating is as old as the world, but not so many documents are still available. It would be interesting to discover how the Egyptians estimated the number of people required to build a pyramid or a temple, or how the Romans planned their engineering activities (aqueducts, bridges, roads or even houses).

Let us start[6] with Leonardo da Vinci (1452–1519): he was commissioned by the city of Genoa to provide a means of setting sale price for vessels based on size and burthen. He then originated a table which gave a "cost per pound" value in ducats to which could be added burdening values plus profit. The tabulation includes[7] true weight-based cost estimating relationships (CERs)!

Thomas Bayes (1702–1761) wrote: "It has occurred to me that mathematics might be applied to the problem of determining *a priori* the cost of say a fine fowling piece, a chronometer or a sextant, … I have observed that time is a factor in the process of manufacture and the more time taken up with the crafting, the more costly the item will be. It should be possible to contrive a model by which money, skill, time spent in labor and the fineness of the article crafted might be related, etc.". One could not say better today!

Brunel (1806–1859), when supervising the construction of his great ship, specified a cost per pound (an elementary CER) for the steel plates, plus a relationship between the rivets count and their cost.

Eiffel (1832–1923), the builder of the famous tower in Paris, following the method of Brunel, used the same kind of relationship to establish the budget of his metallic bridges.

Cyrus W. Field (1819–1892) was interested in laying a submarine telegraph cable between America and Europe. One of the problems was to build the cable (Lord Kelvin: 1824–1907, was contacted for its insulation) which required several innovations. Brunel was also contacted for a cost estimate and suggested to use a number of monetary units per £ of weight. Eventually Field estimated the cable at $0.28/ft or $0.75/£.

The most important development of parametric cost estimating was done during the 20th century, first with the publication (1948) of the MTM, already mentioned, then with the work of pioneers such as Chilton (who generates CERs for different categories of equipments), Wright and Crawford (who studied the cost improvement curves) to mention only a few names. Frank Freiman was the first person to prepare (in the middle of the 1960) what we call here a general model for hardware, and Barry Boehm published his book for cost estimating software in 1981.

### 3.3.3 The Foundations

As strange as it may appear, parametric cost estimating is first based on two beliefs:

1. Costs do not arrive by chance.
2. Then there must be a relationship between the product description and the cost: this is the general principle of causality. Our job, as cost analysts, is to discover it.

Let us discuss for a moment these beliefs. You know quite well that costs do not arrive by chance and you may add this is the reason why the use of such or such machine must be known, as well as the number of workers, the energy consumption,

---

[6] Most of the stories were published by Keith Burbridge in his booklet entitled "*A touch of history*".
[7] CER stands for 'Cost Estimating Relationship'. We will not use these words because many cost estimators forget that a mathematical tool, a model, for forecasting cost is the set of a formula (which they call a CER) plus the scattering around this formula. Consequently a CER cannot be considered as a cost estimating model.

etc. We must therefore be more specific. We can say: starting from a product of which cost is known, if we want to increase its size in order to make a new product, we believe there is a direct relationship between the cost of the new product and the change of size, "direct" meaning there is no need to reestimate the number of workers, the energy consumption, ... unless of course the technology changes. Two important sentences are there:

1. Starting from a product of which cost is known: parametric cost estimating always starts from a knowledge of the cost of "old" product(s). The way these costs are used may differ from one parametric approach to another one (and we will see in Section 3.5 there are two major approaches), but they have to be there: parametric cost estimating is not a substitute to the creation of a database, even if this database can be limited compared to other cost estimating processes.
2. We want to increase its size. This illustrates the basic use of parametric cost estimating (the way it started): the behavior of the cost, when just the product size changes, follows a specific pattern which can easily be seen on a graph and therefore can be translated into a mathematical relationship.

Evidently other factors contribute to the cost changes, as the manufactured quantity, or the quality level or any other functions to be added. At this stage it is important to note that parametric cost estimating starts by taking into account two things:

1. The *product* first: the classical approach looks at the drawings in order to decide what machine will be used and how, whereas parametric cost estimating wants to adopt an "external" point of view. Ideally we would like to start from the functions the product is made to fulfill. But a function can, generally speaking, be fulfilled by different ways, which forces us to have a look at how the function is fulfilled, which can only be done by looking "internally" to the product. But the difference with conventional cost estimating is there: the internal look is, in this approach, the basis for the cost forecast, whereas it is used, in parametric cost estimating, as a complement of the external perspective.
2. And then the *production environment*, from the production organization to the level of automatization.

Let us summarize: the "parametrician" is convinced that the product cost originates from the functions it has to perform. The rest is just there to improve the forecast quality, ...

We can give now a **more precise definition** of parametric cost estimating:

*Parametric cost estimating is a technique which aims at going directly from the known costs of existing products to the cost of a new product, taking into account the change in the technical specifications (as defined by the product user) and in the production environment.*

### 3.3.4 Several Levels of Consistency

The parametrician is convinced that costs do not arrive by chance and can be explained mostly by an "external" point of view. One can also say that this person is convinced that there is some consistency between the products costs and that his/her job is to find it out. This search practically defines the tools he/she will use every day.

# Overview of Cost Estimating

There are several levels of consistencies:

1. Among first the products belonging to the same product family, produced by the same company. This is an elementary consistency.
2. Among all the different products made by this company, even if they fulfill different functions.
3. Among all the products made in the same country, at a given time.
4. Among all the products made in the same country, whatever the time.
5. Among all the products made in the same civilization.

Going from one level to another one requires to add parameters in order to take into account more and more differences.

Everything of course starts at level 1: it will be the subject of Volume 2. Cost estimators should start at this level.

General models then try to go to the other levels, at least 2 and 3, with some hints towards level 4 (the problem being that not so may data are available about historic projects). They will be briefly commented in Part IV.

## 3.3.5 Two Different Types of Models

We clearly distinguish in this book two types of models; we call them "specific" and "general" models.

### Specific Models

These models address level 1. They are built for a "product family" (the word is defined in Volume 2), as homogeneous as possible, by looking at some known products belonging to the same family (the set of these products will be called a sample).

They are strictly built by computation, which means they do not incorporate other information beyond what is included in the sample.

These models are very important for the cost estimator, because they are built from known data, using well-defined procedures. Therefore, unless the data points are very scattered (which probably means they do not belong to the same product family, or that important parameters are not known), their credibility is high.

Due to their importance, the whole Volume 2 is dedicated to them.

### General Models

General models aim at the other levels, at least 2, maybe 3, up to 4 if it were possible.

They obviously do need data, but do not generally use them the same way, because they include other information. The model builder gets these information from his/her experience and thinking, from which some concepts are built; for this reason we often call these models "conceptual models".

As these models are not built by pure computation, the way they are realized cannot be taught. Therefore, we will limit the discussion to the different ways that model builders can follow. This will be the subject of Part III.

# 4 "Elementary" Cost Estimating

## Summary

This chapter is just an introduction to the subject of basic cost estimating at the equipment or product level, the word "product" having to be understood in a very large context.

It starts with the minimum: just one data point is available in the database which is similar enough to the product to be estimated for allowing a direct analogy. It considers the case where no other information is available and the case where we have an idea on how the cost changes with the size.

It then considers the case where two data points are available for making an analogy, investigating the linear and the non-linear approaches.

Eventually it looks to the case where only two data points are available and two parameters are necessary in order to derive a cost figure.

All the concepts can easily be extrapolated. But the real and more interesting question is: "What can be done if we have several data points available and a few parameters (less than the number of data points of course) are necessary?" This will be the subject of Volume 2.

"Elementary" cost estimating investigates what can be done when a very limited number of data points are available for preparing the cost estimate of a new product.

In such a situation, it is obvious that the nature of the product has to be the same: when the number of data points is limited (sometimes to one), we have no way, with the exception of the judgment, to take into account a change in this nature. For the same reasons, the quality level and the production environment must be similar.

As the nature is the same, the only difference between the products in the database and the product to be estimated is the size, which is, as it was said in the previous chapter, one of the most important cost drivers.

## 4.1 Only One Data Point Is Available

### 4.1.1 No Other Information Is Available

In this extreme situation, the only thing which can be done is to linearly extrapolate the cost of the reference product P. The geometric interpretation is given in Figure 4.1. Obviously the size of the product to be estimated must be defined the same way as the size of the reference product.

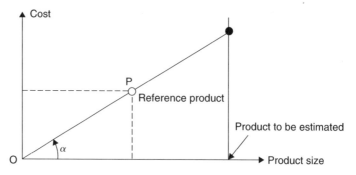

**Figure 4.1** Geometric interpretation: only one data point is available with no other information available.

The product size is here most often quantified functionally (examples: liters/ minute, tons/hour, kilowatt,...) but a physical descriptor, such as the mass, may be used instead.

This method is generally known under the name of "method of the ratio", the ratio being:

$$\text{ratio} = \frac{\text{cost}}{\text{size}} = \text{specific cost}$$

The specific cost, defined as the cost per size unit, is equal to the angle $\alpha$ (more exactly to its tangent) and is *assumed* to be constant for all the products similar to the reference product.

Mathematically it means that, whatever the cost function according to the size Cost(size), you consider that:

$$\text{Cost}(\text{size}) = \text{Cost}(\text{size}_0) \times \frac{\text{size}}{\text{size}_0}$$

where $\text{size}_0$ is the size of the reference equipment.

This method, although very elementary (because there are plenty of evidences that cost does not vary linearly with the size), is still largely used. In the mechanical industry the specific cost is known as the cost per kilogram (or the cost per pound) and is assumed to be constant. As it is well known that it is not correct, one may wonder why it is so largely used; we believe there are several reasons:

- It is true that, with only one data point and no other information are available, it is the only practical solution.
- It is easy to use: the human mind handles easily linear relationships, not so easily more complex ones. This explains why most managers use the specific cost as a first guess of the cost.
- The culture largely favors the belief that cost varies linearly with the size. From our early purchases, we learnt, day after day, that prices depend linearly on the size: current prices are always labeled per kilogram or per liter or per meter and people easily make a confusion between price and cost. Why are prices always labeled in unit per size, at least for the consumer (because the industrial buyer knows very well that prices depend on the ordered quantity)? Just because it is

very convenient: suppose the price of apples were €1/kg for 2 kg, €0.95/kg for 3 kg, €0.91/kg for 4 kg; current purchase would be a headache! Therefore fixing a constant price per kilogram – which must be computed from cost for an average purchase – is much more practical. The same is true for everything, including the fare given to the taxi driver.

However, as soon as the purchase becomes important, consumers know that the price per unit of size decreases with the size. Think for instance at the price of houses: per m² a large house is less expensive than a small house.

Using a ratio should therefore be limited to:

- products of exactly the same nature,
- products of about similar sizes.

Such ratios are very frequently used in the construction industry. For instance Batiprix (publications du Moniteur) is a long list of such ratios usable for building.

It must be recognized that many people know very well that the linear relationship is only an approximation: so, they say, use this ratio for a size between $x$ and $y$, another one for a size between $y$ and $z$, and so on. This is already better, but if you know that the linearity is only valid in a limited range, why not using another type of relationship which will give better results? Now that everybody has a computer, the simplicity of a linear relationship cannot be an excuse anymore.

### 4.1.2 Other Information Is Available

Other information is the shape of the relationship between cost and size. The relationship generally takes the form of a formula such as:

$$\text{cost} = A \times \text{size}^s$$

Such a formula can be used even if you have no reference data point. However using a data point improves the credibility of the estimate: the exponent $s$ is probably (about) valid for a whole industry, but the coefficient $A$ may vary from one company to another one. The reference point allows computing this coefficient.

Figure 4.2 gives a geometric interpretation of this relationship, where only $s$ has to be known.

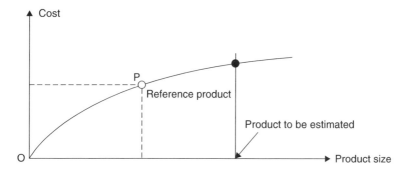

**Figure 4.2** Geometric interpretation: only one data point is available with other information available.

**Table 4.1** Example of exponents.

| Equipment nature | Type | Size unit | Exponent |
|---|---|---|---|
| Agitators | Propellers | hp | 0.50 |
|  | Turbine | hp | 0.30 |
| Air compressor |  | ft$^3$/min | 0.28 |
| Air drier |  | ft$^3$/min | 0.56 |
| Centrifuges | Horizontal basket | Diameter (in.) | 1.25 |
|  | Vertical basket | Diameter (in.) | 1.00 |
| Crane |  | Tons | 0.60 |
| Elevator |  | Height (ft) | 0.32 |
| Hydraulic press |  | ft$^2$ | 0.95 |

What can be the value of $s$? This value depends on the product nature and on the way the size is defined: you will obviously not get the same $s$ for the size given in kilogram and in liters/minute. A lot of values have been published in the literature (see for instance Ref. [18]); Table 4.1 (from K.M. Guthrie) presents, for illustration purposes, a few values used in the petro-chemical industry. Needless to say, the cost found by using this information is more reliable than the cost found by using a ratio.

## A Few Words About the Exponents for Hardware

Table 4.1 gives some examples of the exponent. A lot of work has been done about this exponent in the petro-chemical industry. However most of this work uses different size units, which makes the exponents impossible to compare: if, for the same type of equipment, one person quantifies the size by the length and another one by the volume, you can expect a much higher exponent on the length than on the volume. This position is easy to understand because it allows going directly from the specifications to the cost. However our purpose is different here: we want to compare exponents and therefore we need a common size unit.

The only size unit which is universal for hardware is the mass: any hardware has a mass and this mass is unambiguous. This is not the same for the volume for which several volumes can be defined for the same equipment; compared to the volume, the mass has however a drawback: it depends on the material used for manufacturing it, whereas the volume does not. Nevertheless the mass is a convenient start.

Exponent $s$ now refers to the change in cost when the size (in mass) of an equipment changes. How do we change the size of an equipment? There are two basic solutions:

1. You can increase the size of an equipment by having several identical equipments working together. A good example is given by the batteries: if you want to increase the power storage, you can use several identical batteries in parallel.
2. You can also increase the size of an equipment by expanding each part of it: if you need an electrical transformer for 100 kW, you are not going to use a set of 10 transformers of 10 kW. You will prefer, starting from a 10 kW transformer, enlarging each part of it until it is able of 100 kW. If you need to transport 50 people you will manufacture a bus instead of 10 automobiles.

This discussion is extremely important when manufacturing cost for hardware is considered; it has nothing to do about the "complexity" of the product.

When you increase the size of a product you must know how the enlargement is done

If you just "repeat" the same component several times (a good example is also a ladder), cost grows about linearly with the size ($s \approx 1$); if you enlarge each element, it does not. Pay attention to this point when using a general model: such a model does not ask you this type of question, just because it assumes that the answer is the second one.

How does the cost grow when each part of a product is enlarged in order to make a more "powerful" product? Let us illustrate by an example. Suppose we manufacture a part whose mass is 10 kg and suppose we have the following decomposition:

raw material: €5
machining: €100
the product cost is then €105.

Suppose now we multiply all the dimensions of the product by 2; the mass becomes $2^3 \times 10 = 80$ kg. What happens to the cost? We can make the following analysis:

- raw material: €40, assuming the procurement cost does not change for such a small quantity;
- machining is not multiplied by 8 because what is machined is the surface and the surface increases only by $2^2 = 4$. Machining now costs €400;
- the product cost now becomes €440.

One can observe that $440/105 = 8^{0.69}$ which means that when the mass is multiplied by 8, then the cost grows by only 4.19. In this case $s = 0.69$.

This does not prove anything: it just illustrates the fact that cost includes components which does not grow at the same pace. Practically for equipments which grows "naturally", which means that only each part grows at the same rate, one generally observe an average value of $s = 2/3$ for ground based equipments.

Now we can conjecture that:

- If, when an equipment grows, its complexity increases (for instance due to the fact that the number of parts does increase), then the exponent takes a value larger than 2/3.
- If however its complexity decreases, then the exponent will be lower than 2/3.

Therefore the value of the exponent, when the size is increased by increasing each element of the product, is related to the "added value" (which gives a first quantification of the "complexity" of a product): for an "average" added value, you get an exponent in the vicinity of 2/3. If the added value in the reference equipment is small, then the exponent is less than 2/3; it becomes larger than 2/3 if the opposite is true.

Once again the way the product is enlarged must be considered and this is the real basis of cost estimating. Let us illustrate that on a very simple example. Suppose you have to buy a chain for carrying a load:

- If you increase the length of the chain, keeping its capacity constant, then the chain is sold according to its length, at so many €/meter.
- If you want to increase the capacity of the chain (quantified by its load in decaNewton), then the price follows the curve illustrated in Figure 4.3. As the added value is small when the capacity of the chain is increased, then the exponent is smaller than 2/3; here about 0.5.

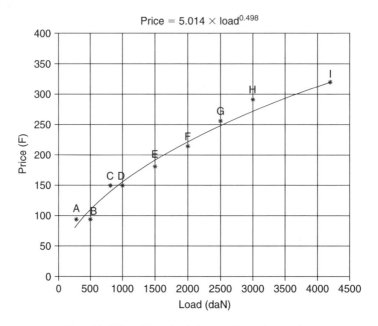

**Figure 4.3** Price of 1 m of a chain according to its capacity.

One element goes in favor of this hypothesis: in the aeronautical and space industries, one generally uses an exponent in the vicinity of 5/6 or 0.83; when a product is enlarged, the added value increases more than in proportion of the mass due to the fact that the designer tries to reduce the product mass as much as he/she can.

Another example is given by electronic equipment for which exponents are in the vicinity of 1. Some cost estimators, extrapolating the previous "law" contend that comes from the fact that the complexity of the equipment is large. We do not agree with this statement: for electronic items we are turning back to the first way of increasing the capacity of an equipment by adding several times the same components. If you want to increase the memory of a circuit you do not enlarge each component, you just add new ones. It is therefore normal that for electronics, the exponent grows[1] towards 1.

These considerations, although they have not the strength of a physical law, may help a lot to get a realistic order of magnitude of equipment even from a single data point.

Note that they are not valid for software: the cost of software increases faster than its size (measured in terms of delivered source instructions). There are also products of which complexity increases very fast, such as the polishing of mirrors: then the exponent is higher than 1.

The fact that for hardware cost increases, for a given type of equipment, with an exponent smaller than 1 explain why it is reasonable to enlarge equipments: there is a benefit to use large equipments rather than a lot of small ones. An aircraft able to transport 300 passengers will be less expensive than two aircrafts able of 150 persons: there is an "economy of scale".

---

[1] If you consider the whole electronic equipment, it includes a container of which cost grows at a smaller rate. Both things being taken into account, you may expect an exponent slightly smaller than 1.

As this is not the case for software, programmers are encouraged to split their software in small pieces; this is effectively the trend.

## 4.2 Two Data Points Are Available

The cost and the size of two products, let us call them P and Q, are now known.

### 4.2.1 No Other Information Is Available

Several computations are possible. People who like the ratios compute the ratios given by both products P and Q and take the average value in order to get a ratio usable for the product to be estimated. The geometric interpretation is given in Figure 4.3 by the mixed line OR of which slope is intermediate between OP and OQ. This gives a first cost estimate.

The second solution is to consider that the line joining P and Q gives a formula for computing the cost of the new product which will be extrapolated or interpolated from P and Q. Let us remind the reader who is not familiar with the subject that the formula giving the cost value, Cost as a function of size, is given, when the costs $C_P$ and $C_Q$, as well as the sizes $S_P$ and $S_Q$ are known, is given by:

$$\frac{\text{Cost} - C_P}{C_Q - C_P} = \frac{\text{size} - S_P}{S_Q - S_P}$$

Let us return to the function Cost(size). What we are doing here is using the Taylor approximation in the vicinity of point P: this approximation is written as follows:

$$\text{Cost(size)} = \text{Cost(size}_P) + \frac{d\text{Cost(size}_P)}{d\text{size}} \times (\text{size} - \text{size}_P) = C_P + \frac{d\text{Cost}(S_P)}{d\text{size}} \times (\text{size} - S_P)$$

where the derivative is approximated by

$$\frac{d\text{Cost}(S_P)}{d\text{size}} \approx \frac{C_Q - C_P}{S_Q - S_P} = \frac{\Delta C}{\Delta S}.$$

The facts that the Taylor approximation is limited to the first order and that the derivative is only approximated mean that this procedure should be limited to the vicinity of the data point P.

The geometric interpretation is given by the dotted line PQ on Figure 4.4. This line is not a ratio anymore because it does not go through point O. It can be considered as a first attempt of building a cost estimating relationship (CER) although it is not really one because the quality of the estimate cannot be assessed; one of its merits is to clearly show that cost does not vary in direct proportion of the size, as the slope of PQ is lower than the one of OR. This produces a second estimate.

Both estimates are not very different if the size of the new product lies between P and Q; but, as the figure clearly shows it, the difference becomes large if it does not.

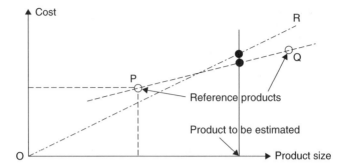

**Figure 4.4** Geometric interpretation: two data points are available with no other information available.

Between these two estimates, which one must be chosen when you do not know how the change in size is achieved? It depends really on the level of confidence you have about P and Q. If this level of confidence is limited, using an average value between the ratios is probably the best choice; if, on the contrary, this level of confidence is high, using line PQ is certainly better.

## A Word of Caution

Be very careful about using a fix ratio (line OR) instead of line PQ: if you quote your proposal according to line OR whereas the correct line is PQ, *you are loosing all the times*, as it is explained on Figure 4.5.

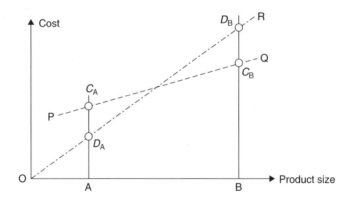

**Figure 4.5** The danger of quoting on line OR.

Let us start with a proposal you make for a product of which size is A. The true cost is given by $C_A$, but you quote value $D_A$: at this price you are cheaper than your competitors and you will get the contract. But you will loose money on it.

Suppose now you make another proposal for a product of which size is B. The true cost is given by $C_B$, but you quote $D_B$; at this price you are more expensive than your competitors: you will not get the contract!

## 4.2.2 Other Information Is Available

The other information can simply be: we think that the law with an exponent would describe better the way cost changes with the size. There is no need to know the exponent – with two data points, one can find both $A$ and $s$ in the formula:

$$\text{cost} = A \times \text{size}^s$$

whatever the way the size is defined. The problem is not difficult to solve – taking the logarithm of both sides, one can write:

$$\log \text{cost} = \log A + s \times \log \text{size}$$

where $\log A$ (and therefore $A$) and $s$ can easily be found out.

The geometrical interpretation is given on Figure 4.6.

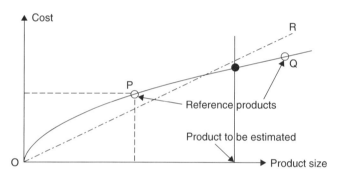

**Figure 4.6**  Using an exponent.

## 4.3 Two Data Points Are Available and Two Parameters Must be Considered

Suppose you think that two quantitative parameters influence the cost, for instance the flow (in liters/second) and the pressure (in Pascal). In such a case you obviously need at least two data points and, in such a case, you will *linearly* interpolate between these data points.

Mathematically it means you now consider that the cost function is a function of two parameters $x_1$ and $x_2$ and you intend to use the Taylor approximation with two variables (Figure 4.7):

$$\text{Cost}(x_1, x_2) = \text{Cost}(x_{1,P}, x_{2,P}) + \frac{\partial \text{Cost}}{\partial x_1} \times (x_1 - x_{1,P}) + \frac{\partial \text{Cost}}{\partial x_2} \times (x_2 - x_{2,P})$$

where the partial derivatives are approximated by a ratio between the data.

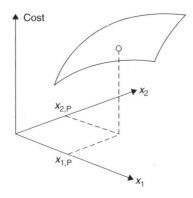

**Figure 4.7** Using two data points and two parameters.

## Example

Let us consider[2] the following data about the production costs of hydraulic pumps for aircrafts:

| No. | Cost (€$_{2004}$) | Production rate (per year) | Flow (l/min) |
|---|---|---|---|
| 1  | 810.25  | 300 | 50  |
| 2  | 517.98  | 300 | 23  |
| 3  | 764.63  | 300 | 45  |
| 4  | 917.41  | 300 | 62  |
| 5  | 856.42  | 300 | 55  |
| 6  | 620.48  | 100 | 20  |
| 7  | 923.10  | 100 | 40  |
| 8  | 672.25  | 100 | 23  |
| 9  | 463.58  | 100 | 12  |
| 10 | 1165.79 | 100 | 60  |
| 11 | 1649.76 | 100 | 110 |
| 12 | 1274.14 | 100 | 70  |
| 13 | 417.50  | 100 | 10  |
| 14 | 583.79  | 100 | 18  |

Obviously the cost depends on both the production rate ($x_1$) and the flow ($x_2$). Later on we will investigate how to take into account all the data in order to prepare a cost estimate for a new pump able of 80 l/min produced at a rate of 150 units per year. Presently we just try to get a first guess with the only tool which is available: the linear interpolation.

The first thing which can be done is to represent the data on a graph (we will insist a lot in this book on the use of the graphs, as they constitute the best way to grasp a lot of information). It is clear in Figure 4.8 that the cost changes about the same way with the flow when the flow changes (this is logic). For estimating the partial derivative $(\partial \text{Cost})/(\partial x_2)$ we can use the data points 11 and 12. We find:

$$\frac{\partial \text{Cost}}{\partial x_2} \approx \frac{\Delta \text{Cost}}{\Delta x_2} = \frac{1649.76 - 1274.12}{110 - 70} = 9.39 \, \text{€}/\text{l}/\text{min}$$

---

[2] This exercise was prepared by Gilles Turré (PSA), who is the author of Chapter 9.

# "Elementary" Cost Estimating

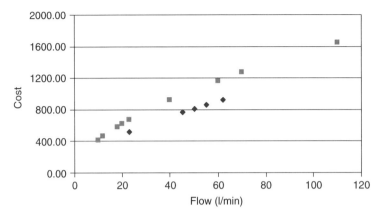

**Figure 4.8** Cost of pumps according to the flow.

In a similar way we can find the partial derivative according to the production rate $\partial \text{Cost}/\partial x_1$ by using data points 4 and 10 (the flow is slightly different between these two points, but it would be a second order correction):

$$\frac{\partial \text{Cost}}{\partial x_1} \approx \frac{\Delta \text{Cost}}{\Delta x_1} = \frac{1165.79 - 917.41}{100 - 300} = -1.24 \text{ €/unit}$$

The minus sign is correct: the cost decreases when the production rate increases (this is also logic).

Eventually, using this linear approximation, will give, starting from data point 12, an estimate of €1781.19 for the cost of the new product:

$$\text{cost} = 1274.14 - 1.24 \times 50 + 9.39 \times 10 = 1306.04 \text{ €}$$

This figure is only an approximation, because, as it clearly appears from the graph:

1. the cost change with the flow is not linear,
2. the change with the production rate depends on the flow.

A slightly more complex model, taking into account these changes, would therefore be necessary and more correct: it would increase the credibility of the estimate.

This is exactly the logic of the parametric cost estimating technique: using all the data which are available to base the estimates on a solid foundation. This will be introduced in the next chapter and fully developed in Volume 2 – for the specific models – and Part III for the general models.

# 5 An Introduction to Parametrics for the Beginner

## Summary

This chapter constitutes an introduction to the parametric way of cost estimating. More exactly it is an introduction to one side of this way, the side, called specific parametrics, which consists in extracting useful information from a database built from data belonging to a specific product family: the adjective "specific" employed with the word parametric means that the useful information is related to one product family and cannot be used for any other family.

The useful information which is extracted from the database takes the form of a formula, plus the scattering of the residual information around this formula.

This chapter illustrates one example, the process by which the formula can be built. Starting from the database itself, it first recommends analyzing the data in order to find out any potential problem. As this chapter is only an introduction to the subject, it does not enter into the detailed analysis and just mention a few points. A complete analysis is carried out in Volume 2.

Once this is done, it shows that several alternatives must be tested before a satisfactory solution is found, the lesson being that the data do not reveal their structure very easily and that, consequently, the cost analyst should try different solutions. The first solution which is generally tried starts with the utilization of one characteristic only, and the use of a functional characteristic is recommended. No algorithm is given in this introduction (these are described in Volume 2): only the results are exposed and visualized on graphs. Other characteristics, sometimes built by the cost analyst from the characteristics which are present in the database, are successively tested.

Once the formula has been successfully built, its quality must be assessed and quantified. In this introduction, only two quantifications are proposed, which give a good idea of this quality.

Once all that is done, the formula plus the result of quality tests are used to estimate the cost of a new object belonging to the same product family. A short discussion precedes the computation, in order to make the reader aware of the logic behind the use of the formula in the forecasting process: this is not immediately obvious, but must be understood in order to properly use the technique.

This chapter concludes by reminding the most important points which were mentioned in this introduction.

Let us assume, it is a pure exercise, that your company would like to know how expensive it could be to buy a commercial aircrafts defined by its capacity in terms of number of seats and range. You have one hour to give an answer* ... .

## 5.1 The Data

You cannot guess and have to base your answer on observed data. Fortunately enough – you have known from the past it is always a good idea to make a record on any cost figure you can find in the literature – you are interested in commercial airplanes and succeeded to gather some data in commercial brochures, reports, etc. The data are the following ones (Table 5.1).

**Table 5.1** Cost and technical information about some commercial aircrafts.

| Name | Cost (M$$_{87}$) | Quantity | Seats | Range (km) | Service year | Empty mass (ton) |
|---|---|---|---|---|---|---|
| ATP | 10.9 | 50 | 64 | 1 700 | 1987 | 15.30 |
| ATR42 | 8.4 | 200 | 46 | 1 800 | 1985 | 10.90 |
| ATR72 | 10.5 | 100 | 66 | 1 800 | 1989 | 13.00 |
| CN 235 | 7.1 | 100 | 44 | 900 | 1986 | 8.20 |
| DHC8-100 | 7.0 | 300 | 36 | 1 500 | 1984 | 10.30 |
| DHC8-300 | 9.5 | 200 | 50 | 1 480 | 1989 | 11.70 |
| EMB 120 | 6.0 | 100 | 30 | 1 000 | 1985 | 7.60 |
| F 50 | 9.5 | 500 | 54 | 2 000 | 1987 | 13.00 |
| SF 340 | 6.5 | 400 | 35 | 1 200 | 1984 | 8.40 |
| 737-200 | 21.0 | 700 | 109 | 3 250 | 1968 | 27.50 |
| 737-300 | 26.0 | 700 | 124 | 3 800 | 1984 | 33.20 |
| 737-400 | 30.0 | 300 | 144 | 3 830 | 1988 | 36.20 |
| 737-500 | 22.0 | 300 | 108 | 3 000 | 1990 | 31.23 |
| 747-200 B | 100.0 | 500 | 429 | 10 790 | 1970 | 173.50 |
| 747-300 | 106.0 | 500 | 475 | 10 150 | 1983 | 178.00 |
| 747-400 | 116.0 | 50 | 475 | 12 150 | 1989 | 183.00 |
| 757-200 | 40.0 | 600 | 187 | 200 | 1983 | 57.20 |
| 767-200 | 51.0 | 200 | 211 | 6 900 | 1982 | 80.60 |
| 767-200 ER | 56.0 | 100 | 211 | 9 870 | 1984 | 83.70 |
| 767-300 | 61.0 | 200 | 258 | 6 950 | 1986 | 87.60 |
| 767-300 ER | 66.0 | 200 | 258 | 10 270 | 1988 | 93.10 |
| 767-400 | 72.0 | 300 | 287 | 8 810 | 1991 | 95.40 |
| A300-600 | 62.0 | 150 | 273 | 6 500 | 1984 | 90.80 |
| A300-600 R | 64.0 | 150 | 273 | 7 240 | 1987 | 90.30 |
| A310-200 | 53.0 | 100 | 220 | 6 900 | 1983 | 79.70 |
| A310-300 | 55.0 | 100 | 220 | 8 000 | 1985 | 82.60 |
| A320 | 34.0 | 300 | 150 | 5 000 | 1988 | 41.20 |
| A340-200 | 85.0 | 10 | 294 | 12 360 | 1992 | 122.00 |
| DC10-10 | 56.0 | 150 | 288 | 6 000 | 1972 | 109.70 |
| DC10-30 | 68.0 | 150 | 288 | 9 850 | 1972 | 122.30 |
| MD-11 | 88.0 | 15 | 332 | 11 270 | 1990 | 133.10 |
| MD-11 MR | 83.0 | 15 | 343 | 7 300 | 1990 | 129.50 |
| MD-11 ER | 83.0 | 15 | 288 | 12 860 | 1990 | 129.10 |
| MD-11 S | 100.0 | 15 | 420 | 9 300 | 1995 | 145.00 |
| MD-81 | 27.5 | 70 | 144 | 2 775 | 1980 | 35.60 |
| MD-82 | 28.5 | 70 | 144 | 3 700 | 1982 | 35.60 |
| MD-83 | 29.5 | 70 | 144 | 4 815 | 1985 | 36.50 |

* all the computations and graphs of this chapter were prepared with EstimLab

For each airplane you get:

- The name of the airplane.
- Its unit price in M$$_{87}$. As you collected these data mainly from commercial brochures, this information is certainly not a cost and probably not the real selling price, as manufacturers often offer discounts on the "official" price; but it is the only information you have. As you do not know the pricing strategy of the various companies, you might expect some scattering of the information on one hand (each company has its own strategy), some consistency on the other hand (companies, when deciding on a selling price, have to take into account the prices proposed by competitors). Both effects may cancel each other, but you do not know, and that will put a limit on the conclusions you may draw from the analysis you want to perform. Let us assume the prices were normalized in terms of economic conditions (same currency, value at the same date).
- The quantity sold.
- The number of seats offered to the public. This number of seats is probably an average (but you do not know) as this number depends on the specific arrangements asked for by the buyer.
- The range (in km).
- The year the airplane started to be sold.
- The empty mass (in metric tons).

You could of course use the procedures developed in the previous chapter, but in such a case you will base your estimate on two data points only. As the cost data may be scattered (you do not know, but you guess so), you consider that you will be much more confident about the estimate if you can base it on all the data rather than on two data points only.

You heard about "parametric"[1] cost estimating and decide, because you do not see any other way to extract a reliable answer from such a collection of data, to test this technique on this particular assignment. You know that there are two very different techniques in this business, called "specific modelization" and "using general models". You do not have a general model and know it takes years to build one; you could certainly buy one, but, at the present time, you have neither the need for such a powerful tool nor the time to get trained to it. So you turn toward the first technique.

You know the basic idea of this technique: from the data you have, you create a relationship telling that, *for this set of data*, the cost is a function of some of the technical characteristics of the airplanes. This function, we will call it a "formula", will be expressed in mathematical terms, from which the cost of a new airplane can easily be computed. This is exactly what you are looking for!

But immediately several questions come to your mind: Among the five characteristics, which one should I select for building my formula? Should I limit the formula to the use of one characteristic only or would it be better to include several ones? How can I quantify what I win by using several characteristics instead of one? What is going to be the accuracy of the cost estimate I will compute (you are pretty sure that your boss will ask this type of question)?

This chapter, which is an introduction to the use of parametric cost estimating, is a first answer to these questions.

---

[1] The name comes from the fact that cost analysts often call "parameters" the technical characteristics they use.

## 5.2 Analysis of the Data

In the past cost analysts, using only one characteristic easily displayed on a graph the data, the characteristic, being set on the "$x$"-axis, the cost on the "$y$"-axis. From that they got a good representation of the data and could easily visualize how scattered they were, if a reasonable trend could be perceived in order to predict the cost from this characteristic, if this trend was linear or not, was continuous or not (a break at some value could come from a change in the manufacturing technology), etc. They could even draw with a pencil a curve which passed in the middle of the data and get the estimate they were looking for by just looking at this curve.

This is still possible and we will use this technique. But the wish to use several parameters in order to tentatively improve the quality of the estimate forces us to use algorithms.

The problem with the algorithms is that they will always give an answer, except in rare circumstances. You will not know if the answer is reasonable or not until you compute the value of some tests; and if the tests do not respond favorably, you do not know why. You should therefore never compute a relationship (a formula) between the variables without analyzing the data first.

Analysis of data when, as it is the case here, several variables are involved is a difficult task because you cannot easily represent the data. The eye is a powerful instrument for gathering a lot of information when they are displayed on a graph (it is rather poor to analyze data in tables), but we cannot see in a multi-dimensional space. Some techniques have been invented for overcoming this limitation, the simplest one being the "star diagram" which is represented in Figure 5.1.

This diagram is rather encouraging: the data are a little bit scattered, which is rather normal. The scattering is higher on the large airplanes, which could be expected to.

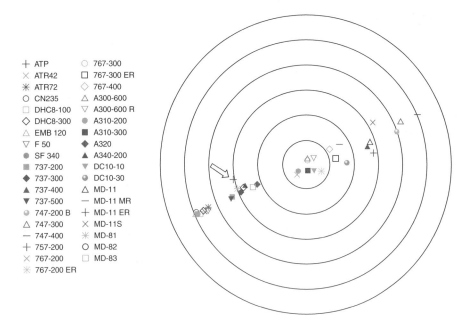

**Figure 5.1** The star diagram.

The interesting things are that no data point (except the 757-200, marked by an arrow) looks abnormal and no "sub-families" appear, as the data points look aligned. Therefore we expect that a good relationship can probably be derived from them.

The discussion of the other visualization techniques is postponed to Volume 2. In supplement to these visual aids, algorithms have been developed for data analysis: their purpose is to compute some tests and to present the results, warning the user about potential problems. Their presentation is also postponed to the same volume, because you have to interpret the results and this interpretation requires some explanations.

The only thing you can do here is to have a quick look at the data; it shows an obvious anomaly: the range of the 757-200 airplane appears far too small: it must be a typing mistake. This line will have to be deleted in the computations. Another anomaly is possible: the year of service of the 737-200 should be verified and maybe corrected (another typing mistake?).

## 5.3 What Are the Relationships Between the Price and the Technical Variables?

The first thing you have to do now is to look for interesting relationships between the cost and the parameters, plus maybe between the parameters themselves and this for two reasons:

1. It would be a plus if you can present to your boss, besides a cost estimate, an estimate of the mass of the airplane.
2. If two parameters quantify about exactly the same thing (such as the mass and the volume for instance: both give a good quantification of the airplane size) there is no need[2] to keep both of them.

This section shows that several relationships may be examined.

### 5.3.1 Relationships Between Price and One-Functional Characteristic

If only one parameter is used, this parameter has to be strongly connected to the size of the "object" you want an estimate for. This is rather obvious: when you buy an apartment, the first thing you think of is the floor surface, when you buy an electrical engine you decide on the power you need, etc.

For a mechanical object two "descriptors" of the size can be used: its physical (such as the mass or the volume) or functional characteristic(s). Functional characteristics should generally be preferred: after all when you buy an airplane you are not buying so many kilograms of an airplane, but so many seats to be transported on so many kilometers. However for estimating the cost of parts, it is often very difficult to quantify their characteristic, whereas the mass is easy to get: this explains why the mass has still the favor of several cost analysts and why we have to look at it.

---

[2] Plus the fact that it can seriously damage the quality of the relationship. But this is another story which will be dealt with in Volume 2.

## Using the Number of Seats as a Functional Characteristic

The first relationship which is interesting from the potential buyer's point of view is, we may think, the relationship between the price and the number of seats (Figure 5.2). In order to investigate this relationship, the first approach is to chart the data: we scale the horizontal axis, the "$x$"-axis, according to the number of seats, and the vertical axis, the "$y$"-axis according to the price.

The chart is encouraging! Number of seats and price are obviously "connected" (and this is logic!); in statistical language, we say both variables are "correlated" which means that, in this case, they grow together. You may note at this stage that, despite this correlation, there is some scattering of the data around what we can call a general trend: for instance the airplane DC10-10 seems to have a low price for its number of seats compared to the other airplanes in its vicinity. Do not draw any conclusion at this stage: the scattering may well be an inherent scattering (due to the price policy of the manufacturer for instance), or it may be due to any other variable not taken into account yet.

This chart is also interesting from another point of view: it shows that the relationship between the number of seats and the price is not really linear: the shape looks like an "S". The cost grows slowly at the beginning, increases then more rapidly until the rate slightly diminishes. The conclusions are at this stage:

1. The linear relationship, so often considered by cost analysts, often without any justification, is only a first approximation of the true relationship.
2. Selecting a type of relationship (linear or non-linear) is a decision made by the cost analyst.
3. We must be able to compute with non-linear relationships. This will be done in Volume 2.

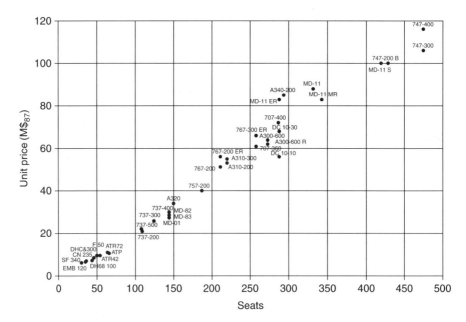

**Figure 5.2** Relationship between number of seats and price (M$_{87}$).

An Introduction to Parametrics for the Beginner

But you presently consider that, due to the limited time, you cannot investigate non-linear relationships and therefore you decide to limit your investigation to linear ones.

You may first want to quantify the visual perception of the correlation you perceive. There are several ways to put a value on that (they are described in Volume 2). The most frequently used was defined by Bravais and Pearson, and is therefore labeled the "Bravais–Pearson correlation coefficient"; this coefficient, looking for a linear relationship between the variables, takes a value between 0 (no correlation at all) and 1 (perfect correlation). Here it takes a value of 0.987, which is quite good!

We spoke about a trend between the number of seats and the price, and decided to investigate linear relationships. You may therefore very well draw with a ruler a straight line through the data (many cost analysts start this way): this line would give you the equation (the formula) of the trend. An easier way is to compute it. How can we compute such a line? In order to do so, you have to decide about three things:

1. The shape of the trend line: in this case we *decided* it will be a straight line. This means we are looking for a relationship such as:

$$\text{price} = a \times \text{seats} + b$$

   where $a$ and $b$, called the coefficients of the formula, are some constants to be determined.
2. How we compute $a$ and $b$? The idea is to search for the straight line which passes as close as possible to all the data points. But what do we mean by close? Generally we say that two points are close together when the distance between them is small. That is fine but, in order to be able to compute, we must define on how distances are measured.
3. We make another decision (other distances will be considered in Volume 2): the distance between one data point, of which coordinates are $\langle \text{seat}_i, \text{price}_i \rangle$ and the straight line will be defined as $[(a \times \text{seats}_i + b) - \text{price}_i]^2$: this distance is the square of the difference between the price value computed by the formula and the real cost.

In order to minimize the sum of these distances – because we want the straight line to pass as close as possible of *all* the data points – the computation gives the following result: $a = 0.251$, $b = -3.589$ which is displayed in Figure 5.3.

The result is not bad. However there is a little problem which is clear from the equation of the trend: for a low number of seats, the price becomes negative! For instance for a four-seat airplane, the formula given upwards leads to a price of $-2.58$ M$! This is obviously a stupid result of which reason appears on the graph: for a number of seats under 40, the data points becomes all over the formula. The curve should go up for low seats values. If you had to estimate the price of a small aircrafts, you should investigate, as previously mentioned, non-linear relationships.

At this time we decide that the number of seats of the aircrafts should not go less than 40.

## Using the Product Number of Seats × Range as a Functional Characteristic

Using only the number of seats as the major functional characteristics seems a bit short: Would it be normal to get the same price for two aircrafts offering the same

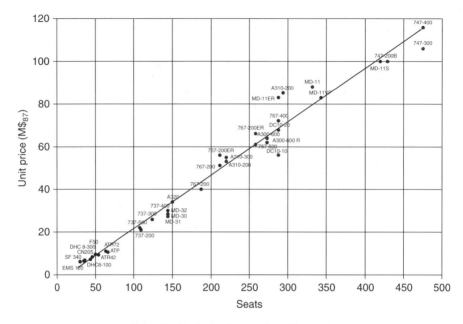

**Figure 5.3** Displaying the equation of the trend.

number of seats for quite different ranges? You do not expect that! The solution we tried up to now, using only the number of seats, may explain the scattering we observe on the graph (Figure 5.4).

A more reasonable solution (while remaining presently with one parameter only) would be to use the number of seats multiplied by the range: it is a combined parameter. The following graph displays the result: the "$x$"-axis represents the product "number of seats × range" (the range was divided by 1000 in order not to get too high figures: the range is then in $10^6$ m or Mm), whereas the "$y$"-axis refers to the price.

There is still a correlation between price and the combined characteristic, especially for the low values of the functional characteristic. However the scattering is larger for the upper values. This may come from the fact that the competition is lower for large aircrafts.

We can also quantify the correlation, using the same Bravais–Pearson correlation coefficient. The value is 0.975, lower than the one we found when using just the number of seats as the only technical characteristic. This appears strange because the correlation does not appear worse on the graph and this result is not expected from a practical point of view.

This comes from the fact that the Bravais–Pearson correlation coefficient is a quantification of the linearity of the relationship. In the present situation, the relationship is not linear and this degrades this correlation coefficient.

Spearman proposed another correlation coefficient: it is based on the ranks and not on the values. This means that, if the "$x$" and "$y$" values are sorted by increasing order, Spearman looks if the ranks grow together or not. In the present case, the Spearman correlation coefficient is equal to 0.992, which is extremely good.

# An Introduction to Parametrics for the Beginner

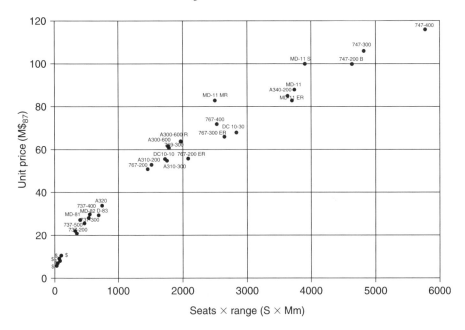

**Figure 5.4** Using the product seats × range as the functional characteristic.

If we look now at the trend, we cannot obviously choose (remember that the selection of the trend shape is your decision) a linear relationship. In order to decide, it is often recommended to display the graph with logarithmic scales. The result appears on Figure 5.5.

Due to this result, it is advised[3] to look for a linear relationship between the logarithm of the price on one hand and the logarithm of the product "seats × range" on the other hand:

$$\log(\text{price}) = c \times \log(\text{seats} \times \text{range}_{Mm}) + d$$

The computation is exactly the same as before and gives:

$$c = 0.587, \ d = -0.15$$

As $0.708 = 10^{-0.15}$, the formula can then be written:

$$\text{price}_{M\$} = 0.708 \times (\text{seats} \times \text{range}_{Mm})^{0.587}$$

The Bravais–Pearson correlation coefficient is now 0.993 between the logs, which is nearly perfect. This looks to be a very satisfactory result.

In order to distinguish this formula from the previous one, it is called "multiplicative" for obvious reasons, the adjective "additive" being reserved for the first one.

---

[3] Here again it clearly appears on the graph that a non-linear relationship should be considered.

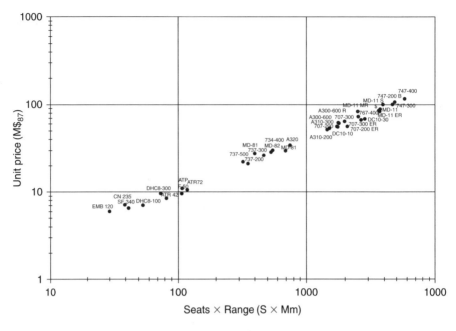

**Figure 5.5** Log–log representation of the price to the "seats × range".

## 5.3.2 Relationships Between Price and Several Characteristics

*Using the Couple Seats and Range Independently of Each Other*

The previous section uses the product "seats × range" as a compound variable. You may ask why using such a compound and not using both characteristics independently?

Now we deal with three variables: price, seats and range. It would be difficult to represent this set in a three-dimensional space (but there are methods [see Chapter 5 in Volume 2] to palliate this difficulty) and we have to replace our eyes by algorithms. These algorithms are not difficult and are explained in this book in many details.

What we can try to do here is to look for a linear relationship between price, seats and range. After all we were rather successful when using a linear relationship between price and seats, and it is a good idea to look if adding the range as another variable can improve the results. Therefore we are looking for a relationship such as:

$$\text{price} = a \times \text{seats} + b \times \text{range} + c$$

The computation is of course a bit more complex than with just the seats, but not really difficult. The following relationships were found for both the additive[4] and the multiplicative formulae:

$$\text{price}_{M\$} = -5.517 + 0.185 \times \text{seats} + 0.003 \times \text{range}_{Mm}$$
$$\text{price}_{M\$} = 0.029 \times \text{seats}^{0.820} \times \text{range}_{Mm}^{0.345}$$

with, in both cases, a correlation coefficient equal to 0.997, which is still a better one!

---

[4] Once again, and for the same reason, the estimated price of a small aircrafts would be negative!

# An Introduction to Parametrics for the Beginner

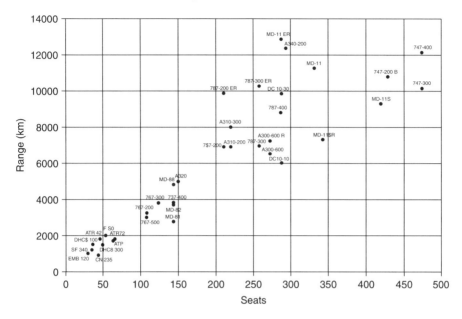

**Figure 5.6** Relationship between number of seats and range.

However something looks strange! When we computed the formula with only the number of seats, we got the following result:

$$\text{price}_{M\$} = -3.589 + 0.251 \times \text{seats}$$

It means if we change the number of seats only, without changing the range, we do not get the same result with both formulae.

We could expect a slight modification but not such a large one. Where does this result comes from? The answer is in the fact that the number of seats and the range are strongly correlated, as Figure 5.6 shows it, even if the scattering is rather large. The correlation is rather severe for small aircrafts, but is limited for large aircrafts. The quantification of this correlation according to our previous authors gives for:

- Bravais–Pearson: 0.889
- Spearman: 0.927

which confirms our feeling: the correlation between both variables is not negligible.

Why does this correlation entails such a difference in the coefficients of the number of seats in the formulae? The answer to this question is rather technical: the reader will find the exact answer in Chapter 6 of Volume 2. For the time being, let us only make this comment: when two parameters express about the same thing, as it is the unexpected case here, the algorithm we use to compute the coefficients "manages" at its best to find the coefficients. It has to compute simultaneously the coefficients for the seats and for the range, whereas these parameters mean about the same thing: it therefore splits at its best the change in the cost on the two variables. You cannot expect a very good split when two variables convey about the same information: more precisely, you may get the right one, but it is a matter of chance. And the split can change dramatically if one cost value only is changed: the formula is unstable.

The conclusion: be extremely careful when using several (quantitative) parameters in the same formula and do not trust too much your feeling; always check the correlation between these parameters. In the present case, one could start by saying: "number of seats and range are obviously not correlated" and check if this hypothesis is correct or not. It may be true technically speaking, but it is wrong from the data we have. Another set of data may produce a different conclusion, but we have to live with our data.

By the way, the correlation is not abnormal: aircrafts manufacturers generally build larger aircrafts for long distance and smaller ones for short distances.

From now on, we will keep the formula based on the product "seats × range", as it gives very good results.

## Does the Service Year Influence the Price?

Starting with this formula, we have to examine if the service year does influence the cost. For investigating this possibility, we create another formula with the product "seats × range" and "service year". First of all we check is there is no correlation between these variables: the correlation coefficients are:

- Bravais–Pearson: 0.068
- Spearman: 0.223.

So there is no significant correlation. Instead of recomputing the formula, it is easier to compute the correlation coefficients between the cost and the service year:

- Bravais–Pearson: 0.013
- Spearman: 0.241.

As these correlation coefficients are negligible, we cannot expect to improve our formula by taking into account the service year. This result is a bit of a surprise because one could expect a change in the price with time (2% a year is generally considered as an average). A conclusion could be that the price is computed by the manufacturer taking into account the productivity gain during the lifetime of the aircrafts: he/she then keeps a constant selling price.

## Does the Quantity Influence the Price?

Let us check first the correlation coefficient between these variables:

- Bravais–Pearson: −0.169
- Spearman: −0.223.

There is – on the average (because the correlation coefficients are very small) – a slight decrease in the price when the sold quantity increases: the sign "−" in front of the coefficients shows an anti-correlation. But the weakness of the values does not promise any significant improvement in the formula which becomes:

$$\text{price}_{M\$} = 0.794 \times (\text{seats} \times \text{range}_{Mm})^{0.581} \times \text{quantity}^{-0.015}$$

with a Bravais–Pearson correlation coefficient equal to 0.993, not really different from the previous one. The reader may notice that the exponent on the product "seats × range" is very close to the one we previously found; this comes from the fact that the coefficient correlation between these parameters is very small (−0.136): adding the variable "quantity" does not really influence the relationship between the price and the product "seats × range".

The reason for such a low change in the price when the quantity increases (when the quantity doubles, the price goes down by only 1%!) probably comes from the fact that the aircrafts manufacturer decides about its prices for the production cost at rank 100 (the cost of the 100th aircrafts produced) and keeps the selling price at the same level whatever the quantity.

There is therefore no reason to keep the quantity in the formula.

### 5.3.3 Relationship Between Price and Mass

There is an expected correlation between these variables.

You do not really buy a airplane by its mass! Nevertheless many cost analysts like to define the size of the products by their mass. This comes from the fact that the mass is generally well known for flying objects, or even for ground mobile items and that it is sometimes difficult to functionally defines products: at least the mass can be compared from one product to another one.

Even if we always prefer to use functional characteristics when making an estimate (simply because it allows to quickly make trade-off analysis or to answer management questions during the decision phase of a project), it is worth considering the relationship between mass and price: Figure 5.7 is very encouraging in

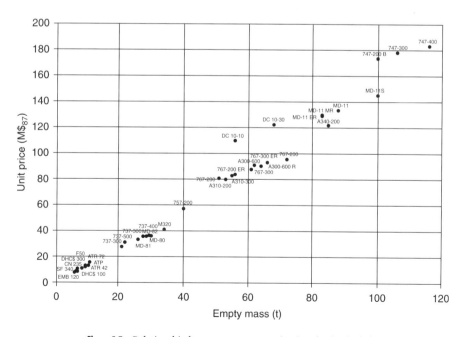

**Figure 5.7** Relationship between empty mass (ton) and price (M$_{87}$).

this respect. The Bravais–Pearson correlation coefficient amounts to 0.988 (and the Spearman to 0.991) which suggests a linear relationship (the "S"-shape – or more precisely here the "J"-shape – between the characteristic and the price is less obvious here but is nevertheless visible on the graph):

$$\text{price}_{M\$} = 4.295 + 0.605 \times \text{mass}_{ton}$$

### 5.3.4 Is There a Correlation Between Mass and Seats × Range?

Technically speaking, there should be a good correlation between both variables. A second reason is that these variables are well correlated to cost. Checking the correlation coefficient we get:

- Bravais–Pearson: 0.975
- Spearman: 0.983.

On log–log scales the correlation looks good, although a "J"-shape is visible (Figure 5.8).

The relationship is easily quantified:

$$\text{mass}_{ton} = 0.729 \times (\text{seats} \times \text{range}_{Mm})^{0.636}$$

with a Bravais–Pearson correlation coefficient now equal to 0.985. This formula allows to compute the mass of an aircrafts when the number of seats and the range are known.

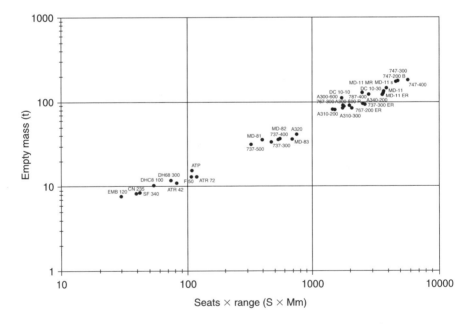

**Figure 5.8** Relationship between the "seats × range" variable and the aircrafts mass.

# 5.4 What Is the Quality of the Formula?

Several things may be looked at once the formula is computed. Presently we will limit our investigations to two points: examination of the residuals, quantification of the quality of the formula.

## 5.4.1 Examining the Residuals

The residuals are the differences between the prices given, also called "observed" prices, for the airplanes and the prices which can be computed for the same aircrafts when applying the formula.

Let us compute (Table 5.2) and display (Figure 5.9) these residuals.

**Table 5.2** List of the residuals.

| Name | Observed price (M$_{87}$) | Estimated price (M$_{87}$) | Residuals |
| --- | --- | --- | --- |
| ATP | 10.9 | 11.2 | −0.257 |
| ATR42 | 8.4 | 9.5 | −1.110 |
| ATR72 | 10.5 | 11.7 | −1.246 |
| CN 235 | 7.1 | 6.2 | 0.923 |
| DHC8-100 | 7.0 | 7.4 | −0.406 |
| DHC8-300 | 9.5 | 8.9 | 0.595 |
| EMB 120 | 6.0 | 5.3 | 0.749 |
| F 50 | 9.5 | 11.1 | −1.609 |
| SF 340 | 6.5 | 6.4 | 0.107 |
| 737-200 | 21.0 | 22.3 | −1.300 |
| 737-300 | 26.0 | 26.3 | −0.300 |
| 737-400 | 30.0 | 28.8 | 1.164 |
| 737-500 | 22.0 | 21.1 | 0.875 |
| 747-200 B | 100.0 | 100.1 | −0.099 |
| 747-300 | 106.0 | 102.5 | 3.488 |
| 747-400 | 116.0 | 113.9 | 2.115 |
| 767-200 | 51.0 | 50.9 | 0.119 |
| 767-200 ER | 56.0 | 62.7 | −6.734 |
| 767-300 | 61.0 | 57.5 | 3.524 |
| 767-300 ER | 66.0 | 72.2 | −6.226 |
| 767-400 | 72.0 | 70.3 | 1.726 |
| A300-600 | 62.0 | 57.1 | 4.874 |
| A300-600 R | 64.0 | 60.8 | 3.155 |
| A310-200 | 53.0 | 52.1 | 0.860 |
| A310-300 | 55.0 | 56.9 | −1.852 |
| A320 | 34.0 | 34.5 | −0.517 |
| A340-200 | 85.0 | 86.9 | −1.884 |
| DC10-10 | 56.0 | 56.2 | −0.245 |
| DC10-30 | 68.0 | 75.2 | −7.167 |
| MD-11 | 88.0 | 88.4 | −0.382 |
| MD-11 MR | 83.0 | 69.9 | 13.126 |
| MD-11 ER | 83.0 | 87.9 | −4.857 |
| MD-11 S | 100.0 | 90.6 | 9.367 |
| MD-81 | 27.5 | 23.9 | 3.618 |
| MD-82 | 28.5 | 28.3 | 0.240 |
| MD-83 | 29.5 | 33.0 | −3.467 |

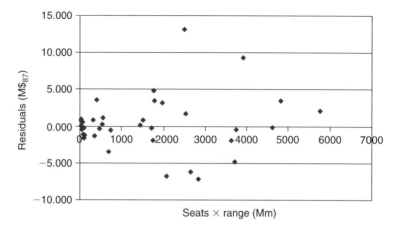

**Figure 5.9** Residuals as a function of the product seats × range.

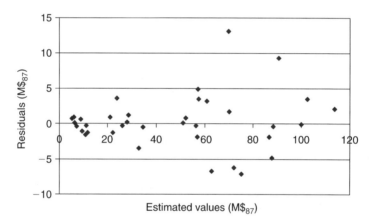

**Figure 5.10** Residuals as a function of the estimated values.

Displaying the residuals as a function of the product "seats × range" does not reveal any singularity or trend, which is good.

Displaying the residuals against the estimated values also does not reveal any problem (Figure 5.10).

We will return to the analysis of these residuals in Chapter 13 of Volume 2.

### 5.4.2 Quantifying the Quality of the Formula

There are two ways, of course linked together, for quantifying the quality of the formula: the coefficient of determination, and the variance of the coefficients.

## The Coefficient of Determination $R^2$

This coefficient gives a broad value about the quality of the formula. Its mathematical form is a bit complex, but its interpretation is easy. Suppose that all the airplanes have the same price (for instance the average price, which is here 47.5 M$\$_{87}$): then the database does not really contain any information, as there is nothing we can extract from it …

The only *usable information* for each aircrafts is therefore the difference between the average price and its given price. As this difference can be positive or negative, and in order to use only positive values, let us define the amount of information[5] available for an aircrafts by the square of this difference.

The total amount of information available in our sample is the sum of these information:

$$\text{Available information in the sample} = \sum_i (\text{given price} - \text{average price})^2$$

Now, when the formula is built, this available information is split into two terms:

1. Information which is included in the formula.
2. Information which is lost: the information corresponds to the squares of the residuals just mentioned.

Now the coefficient of determination is easy to define:

$$R^2 = \frac{\text{information included in the formula}}{\text{available information}} = \frac{1 - \text{information lost}}{\text{available information}}$$

You obviously want as much information included in the formula as possible: therefore you would like to get a $R^2$ as close as 1 as possible. Practically when dealing with cost or price, one can be rather pleased if the $R^2$ is greater than, let us say, 0.85.

With the formula based on the number of seats multiplied by the range, we get $R^2 = 0.986$ which means that 99% of the information is incorporated in the formula and only 1% is lost. This is a very interesting result.

An other easily remembered definition of the $R^2$ (although not really mathematically correct) is:

$R^2$ multiplied by 100 gives the percentage of the available information which is "explained" by the formula.

If you have to explain why you are confident about your estimate, you may use this definition in order to explain that you computed your estimate from a formula of which the $R^2$ has such a value. Everybody will understand it.

## The Variance of the Coefficients

The variance of the coefficients is a statistical property. The idea is the following one: if it were possible to get several other sets of 37 airplanes, we would expect the coefficients to vary from one set to another set. The word "variance", or more exactly its square root called the "standard error", quantifies this variation: a large

---

[5] This is of course not the definition of information given by Shannon. You may use another term if you prefer.

standard error means you should not really trust the value of the coefficients, as they are rather imprecise.

Nevertheless a standard error of 10 on a coefficient of 1000 is not important at all whereas the same standard error on a value of 1 is really embarrassing! So, instead of displaying the variance or the standard error, it is more convenient to display the ratio of the coefficient divided by its standard error. This ratio is generally called the "*t*" ratio.[6] Theoretical considerations (explained in Chapter 15 of Volume 2) show that this "*t*" should be greater than 2 for having some confidence in the formula.

Here we have two "*t*" ratios (Table 5.3).

These results are very satisfactory: the formula we found is trustable.

**Table 5.3** Standard errors of the coefficients.

| Coefficient | Standard error | *t* |
|---|---|---|
| Intercept | 0.086 | 8.35 |
| Seats × range | 0.009 | 62.28 |

## Other Results

One can also have a look at:

- The Fisher–Snedecor test: the average residual here is 70.9 which should be compared to its threshold of 2.84: this test is also very satisfactory.
- The average absolute value of the residuals, which gives an easy way to understand the test of the predictive capacity of the formula: the average residual is here 6.2%, which is not wonderful!

It is interesting to plot the absolute value of the residuals in percent against the cost values (Figure 5.11).

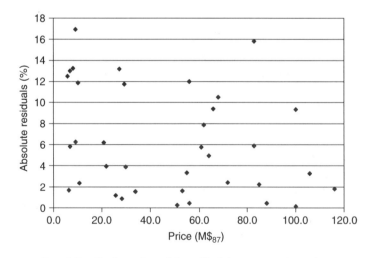

**Figure 5.11** Absolute values of the residuals in percent against prices.

---

[6] This name was given to this ratio by Gosset, who studied it.

This graph shows that, although all the statistical tests ($R^2$, Fisher–Snedecor, ...) are favorable, the results are not as perfect as expected ... This explains the computed absolute value of the average residual of only 6.2%.

In our view, this simple, unsophisticated, easy to understand, value gives a much better idea of what can be expected from the formula than more complex tests (do not forget that the residuals can be positive or negative, as only their absolute values are plotted) because in the cost domain we are used to discuss percentage values and not so much absolute values.

## 5.5 Estimating the Price of a New Aircraft

Now that we have studied the data, found interesting correlations and built the formulae, it is time to try to use them for estimating the price or the mass of a new aircrafts.

But we have a problem: we found a formula for a **sample** containing 37 aircrafts. Are we allowed to use this formula for estimating the price of any other aircrafts?

First of all, and this is rather obvious, the new aircrafts should belong to the same product family (the concept of product family is introduced in Chapter 1 of Volume 2): you are not going to estimate the cost of a new military aircrafts based on commercial aircrafts!

Second we noticed, because we decided to use a linear relationship, that the formula must not be used for small aircrafts. A non-linear relationship should be developed for these aircrafts. The idea is: before using a formula for estimating the cost of a new product, even belonging to the same product family, check if this use is valid or not.

Third, the relationship is valid for the sample and we have no way to demonstrate how it can be used for any other aircrafts, even belonging to the same *product family*. How can we proceed? The idea is then the following one:

1. We consider all the commercial aircrafts which could be built the same way (using the same technologies) as the aircrafts belonging to our sample. The set of all these possible aircrafts is called "the **population**".[7]
2. We consider that the aircrafts belonging to our sample were "randomly" drawn from this population (we know it is not true as we never drew randomly aircrafts from a large population, but this assertion is made for explaining the logic). No particular airplane was selected on purpose from the population.
3. As no aircrafts was specifically chosen for our study, we can legitimately think that the relationships we found in the sample can be applied, with maybe a possible error, to any airplane belonging to the same population, within of course certain limits.
4. Therefore we *decide* that the relationships will be applied to any aircrafts. The only thing we do not know at this stage is the possible error we can make when applying these formulae. This error may have three causes (plus the fact that the

---

[7] The term is standard in statistics: it comes from the fact that statistics were primarily used for studying human populations. Here we speak about populations of aircrafts.

new airplane may not exactly belong to the same product family, but we assume it is not the case):
 - The scattering of the data: it shows that the formula cannot be completely exact.
 - The sample size: it has a limited size; when it was "drawn" from the population, we may have been lucky to get airplanes which fit nicely with the formulae we get. Of course this possibility decreases when the size of the sample increases, but it always exist. Consequently the confidence we may have in the formulae depends on the sample size.
 - The position of the new aircrafts inside the sample. Intuitively we will have more confidence in the cost of a new airplane located in the middle of the sample, than for an airplane located at one end of the sample. At the end of the sample we are never sure that the formula is not going to change, as we saw it previously at the left end of Figure 5.3. Consequently the confidence we may have with the computed value depends on the position of its describing parameters inside the sample.
5. It is then possible to compute the possible error which can be made for the estimate, taking into account all these comments. The result of this computation – developed in Chapter 15 – gives, for an airplane of 600 seats and a range of 12 000 km: 129.8 ± 1.5 M$$_{87}$ at a level of confidence of 90%.

## What Have We Learnt so Far?

1. Never start computing without checking the data first.
2. When looking for a trend between the price and one characteristic, select the shape of the trend – the equation of the curve – that best fit your data. You may start with a linear trend, but, in the domain of cost, linearity is rather infrequent.
3. Pay attention to what happens at the ends of your data range. Quite often when using a functional characteristic, linearity stops at low values.
4. The Bravais–Pearson correlation coefficient is a quantification of the linear correlation between two variables. The Spearman correlation coefficient, because it just looks at the ranks, does not care if the relationship between the variables is linear or not. If the second is higher than the first, this generally means that the relationship is not linear.
5. You are the one who decides the shape of the relationship between the variables. To start with, limit your choice to linear and multiplicative formulae. There are other ones but you can solve at least 70% of your problems with those two.
6. When you have two parameters you consider as technically not correlated because they represent different things, check nevertheless if they are not physically correlated, such as the number of seats and the range, because large airplanes are generally made for long ranges. If it is the case, try to use a compound variable, such as the product of both (other solutions, such as the "Ridge regression", are presented in Volume 2).
7. When using several parameters, always check about the possible correlations, between them: do some parameters express about the same thing? The answer is not always technically obvious: therefore you must study all the couples.
8. Use your eyes as much as possible: graphs are an extremely powerful to convey information to your brain. A whole chapter is dedicated to that in Volume 2.

9. You can learn a lot of things about the market by studying the various correlations between the variables.

## General Conclusion

Computing a formula and using it for cost estimating purposes is a serious business and no improvisation can be accepted. This book was written for you to get the maximum of this technique.

# Part II
## Preparing the Data

## Part Contents

Chapter 6 **Data Collection and Checking**

Chapter 7 **Economics**

Chapter 8 **The Cost Improvement Curves**

Chapter 9 **Plant Capacity and Load**

Chapter 10 **Other Normalizations**

In the cost domain, when we try to forecast what the cost of a new "object" (it can be a product, a software, a building, or anything else), it is becoming more and more difficult to find a similar object we made in the past. It used to be the case, long time ago. But the rate of change has become too fast to make us able to forecast this way. What we try therefore to do is to take several objects and, from the knowledge we have about these objects and their costs, to try, by **comparing** them, to find a solution to our cost estimating problem.

The key word here is "comparing": parametric cost estimating is nothing else that a quantitative approach to extract, from our comparisons, something useful to solve our problem. The purpose of this book is to make you able to extract information on the best possible and efficient way.

We can only compare "comparable" information. The first question you have to ask yourself when you receive a new cost information is: "*May this cost be compared to other costs I have?*". Answering this question may seem obvious and many cost analysts we know do not ask it. However, we think it is one of the most important questions to ask before we start any comparison, whatever it may be. The path we will follow is called "normalization". What we mean by that has two sides:

1. First, you **define** what the *standard situation* should be: we will see the components of this situation. Very briefly, we can say that the costs must be computed the same way, in the same currency, in the same economic conditions, in the same working environment, for the same production rank, at the same time in the product life cycle.
2. Second, and this assumes of course that you know what was the situation that prevailed when the costs you have published, you **adjust** the cost figures in order to compute what the costs would have been if they had occurred in the standard situation. The adjusted costs are said "*normalized*" (to your own standard situation).

*Once you have done this normalization, and this part explains how it can be done, you can start comparing the costs.*

Once the model is built, it is used to compute the cost of something new: of course the cost will be given as if it would occur in the standard situation. It must therefore afterwards be "*denormalized*" in order to have it in the situation you are interested in.

Normalization and denormalization go together, as both faces of the same coin. They will be dealt with in the same chapters.

Another reason for discussing the normalization/denormalization is that it will introduce the reader to the modeling process.

# 6 Data Collection and Checking

## Summary

Data collection is one of the most important step for being able to prepare future cost estimates – as reminded in the previous part – as the only way we, as human beings, are able to forecast the future is by extrapolating the results of past experiences.

A first question which has to be answered is: What data should be collected? In the cost domain it is impossible to "generate" data and we have to live with the data we can get. However, experience shows that most data are incomplete and therefore sometimes unusable. This chapter starts by reminding this fact and insists on the fact that a cost by itself is worth nothing if what is measured in this cost is not properly defined: economics, product description, list of activities, abnormal phenomena, etc.

It also reminds the fact that (and this regards mainly large organizations) internal costs should also be considered, especially if the ratio between external and internal costs changes from one project to another one.

Expert judgments are very interesting pieces of information and should be added, each time it is possible. In this domain one important point is to be able to cope with the experts' subjectivity which is a normal phenomenon. This is generally speaking difficult to detect; one way is to check the experts' consistency; this chapter gives some hints in this respect.

Large organizations (and sometimes small ones) must subcontract projects or parts of projects. Contracts are important data if the price figures are properly displayed which means that:

- The prices hierarchy should be the same for all projects. For instance the same product breakdown structure (PBS) must be used for the different types of projects such as production and development: the basis of the structure should therefore be the product structure.
- Product/activities descriptions should be added: generally product/activities descriptions are not given in parallel to price description, which makes price allocation difficult.
- Details must be given. Most often details are present in the proposals but are not updated in the contracts, especially when negotiation is made at the project level.

This means that the structure should be a two dimensions matrix: one dimension should be the product structure, the second one being the activities. Quite often both structures are mixed: this produces the common work breakdown structure (WBS); this may be sufficient for small projects, but for large projects, we advocate the two dimensions matrix structure.

There is a normal trend towards fix price contracts as opposite to cost plus fee contracts. This is quite understandable for both parties. However, organizations must know that, in doing so, they loose a lot of information; this loss must be compared to the advantages of using these fix price contract.

We already said that the best way the human mind found for forecasting the future is to extrapolate, in the most efficient way, the results of past experiences. There is no other way.

This implies that the data related to past experiences are properly saved and easy to recover.

The cost estimating domain is, in this respect, similar to what happens in human sciences: practically no experiment is possible, as we do it in physics. The only information we can get comes from past experiences.

From our experience, data collection is still the weakest point of the cost forecasting process. Managers should be convinced of its importance and allocate the necessary resources to it. Data collection cannot be an automatic process: data collection does not mean only saving cost figures. A lot of information must generally be saved at the same time in order to make these cost data useful.

If saving cost figures can easily be automated, this is not the same for all the other information, which will be briefly described in the following pages. Consequently data collection should be regarded as a true function in the companies, and an important one: *the future of the company depends partly of it*. Once it is considered that it is a vital function, it has to be organized and funded.

It is an investment which pays: properly saved information saves a lot of time when cost estimates are required. Money invested in this investment will pay back in a short period of time; it will divide the cost of an estimate by factor of at least 3. And talented cost estimators are an expensive resource!

At the same time it improves considerably the credibility of the cost estimates: management knows that the cost figures which are presented are based on reliable sources. If, at the same time, the process of extracting useful information from the database is considered as highly professional, then management knows it can base its decision on them.

## 6.1 Collecting "Raw" Data

Sources of data are, generally speaking, abundant … Sources of reliable data are scarce …

Data collection just means to enter data in what is generally called a "database".

### 6.1.1 What Do We Mean by Data?

For each product entered in a database one should get three types of information:

1. *Product*[1] *description*: What is it? The response to this question may be analyzed in different terms (functional, physical, technology, number of parts, etc.).

---

[1] "Product", as previously explained, is a generic term for what the cost refers to: it can be an object, a software, a building, a tunnel or even an activity.

2. *Activities description*: What activities are included in the cost? That should be as detailed as possible. It should not be forgotten that costs measure activities, not products. We have to establish the bridge between them.
3. *Cost description*: What are the costs? Currency, economic conditions, etc.

If one information is missing in this triplet, the data is worth nothing, or so little.

### 6.1.2 The Point of View of the Cost Analyst

Ideally sources of data are costs observed on past projects in your own organization. Here the cost analyst has potentially access to the costs themselves.

For projects which are partially or totally subcontracted, only prices are known. On one hand it is a more difficult situation, because the relationship between cost and price is not known, and is probably influenced by several variables (amount of competition, present workload, desire to enter a market or not, level of expertise in the subject, etc.) which are not known. On the other hand several proposals are generally received; a careful analysis of these proposals, taking into account some information we may directly get from the companies, may give some useful hints about the costs.

*What Information to Collect?*

There are two different schools:

1. *Some people say*: "let us buy a database and structure it according to our needs. Once this is done let us start collecting data".
2. *Other people say*: "let us start collecting data and let us start with a reasonable, not too expensive, database which requires the less amount of structure. It will be the work of the analysts to dig for data in this partially structured database".

There is some logic in both opinions. The choice between them depends a lot on the amount of information you have to save. If you work with thousands of different products, the first one is obviously the correct solution and you certainly have this database already.

If you work with 10 or even hundred products and are just starting the process of building a database, the second solution is preferable, because you do not know today what kind of data will be really helpful and, times changing, the information you may need tomorrow will not be the same as the one you need today.

Our opinion, if you are starting the process, is to save all the information you may have about a product, including reports, pictures, detailed costs, manufacturer, specifications, duration, incidents, tests, etc. Experience shows that data so collected may be useful in the future.

*About the Internal Costs*

It may happen, in several organizations, that projects are partly done outside the organization (subcontracted) and partly done inside the organization.

Internal costs should be recorded as well as external prices: it is the only way to make projects comparable. Projects which are partly subcontracted will appear less expensive that other ones if internal costs are not added (it is a way which is used in some organizations to make the projects appear cheap). As we are interested to costs to the organization, both costs should be taken into account.

### 6.1.3 A Word of Caution

In many companies we saw, data collection and data analysis are carried out by different persons. What happens then?

People who are in charge of data collection do their best, look at past contracts, ... and enter the data in a database. They do not really know what is to be done with the data: it is not their job.

When a cost analyst starts using the data, you may be sure that some useful information was forgotten, that data are not consistent, that data are aggregated in some way for some products, in another way for other products, etc. The result is that the cost analyst would like to go back to the original data; but this is too late: it will be too time consuming, or the data have already been destroyed.

Therefore our advice is the following one: people who collect data should know about what will be done with the data. They should be basically trained on data analysis and should make themselves a preliminary analysis; they do not have to carry out the complete analysis, but at least to make some analysis. They will therefore discover that some information is missing, that a typing mistake was made, that something was forgotten, that some details should be looked for, that some data are inconsistent, etc.

This preliminary analysis must be made immediately, when the raw data are still there, available. Afterwards it will be too late and too time consuming.

The division of labor should not go too far in the data collection business!

### 6.1.4 What Database?

This comment is dedicated to beginners: start with documents and files, each for each project. When the number of projects goes over two or three 10s and is regularly increasing, it is time to switch to an electronic database.

Your first database should not be too large. The principal qualities[2] are as follows:

- *Easiness to use*: do not spend too much time to manage it, as you have better things to do.
- *Flexibility*: as you do not know yet what you will have to record if your experience is limited.
- *Transferability of data*: if your volume of data becomes large, you have to switch to a more powerful database, the important thing being then not to be obliged to redo everything.

---

[2] EstimLab™ recognizes it is an important function and proposes a simple and easy to use database structure.

The important idea is very simple: "dedicate more time on your data (clarification, analysis) than on managing a database!".

## 6.2 Data Clarification

The purpose of data clarification is to make sure that the data, on the first look, can be comparable. Data clarification precedes data normalization which will be the purpose of the following chapters. The difference between data clarification and data normalization is that the first one is (more or less) subjective, whereas the second one is more objective (which does not mean it can be completely automatized, some judgment having sometimes to be used).

Data clarification has five components:

1. Making sure that the "products" are well defined.
2. Making sure that the activities which are included in the cost are well understood.
3. Making sure that no "over-cost" was charged to the product.
4. Making sure that the costs are properly defined.
5. Making sure that, when expert judgment is required, subjectivity is (as much as possible) eliminated.

### 6.2.1 Products Definition

Cost estimating is based on comparing products to be estimated with known products. As this comparison cannot often be direct, "indirect" comparisons are made. Indirect comparisons mean that we try to extract from known products information what is useful for cost estimating and then to extrapolate it to the product to be estimated.

Whatever the comparison, we need some information about the products themselves. The amount of information which is needed cannot be precisely defined as it depends on the tool which is used for cost estimating: you need much more information to be able to carry out a detailed cost estimate than to use a model; you need more information to use a general model than to use a specific model.

Let us concentrate here on the specific models. A specific model is built and used inside a dedicated "product family" of which definition is given in Chapter 1 of Volume 2. Very briefly a product family is a set of products fulfilling the same functions in about the same way, which means that products can easily be compared inside the same product family.

The interesting feature about dealing with products belonging to the same product family is that we do not need a complete definition of them, but only a description of their differences. The first and the most obvious difference is naturally the products size: this explains why many specific models use one variable only, which is the size (which can be quantified in *physical terms*, such as the mass or the volume, or in *functional terms*, such as the power or the capacity).

But other differences may exist such as an auxiliary function which is added on some products and not on other ones, the material for hardware products, the technology for electronic products, the rock to be excavated for tunnels, the location for a building, etc. These differences must be known.

From our experience the product definition, we should better say the lack of proper definition, is the weak point of most databases. This is a rather strange thing to say for products manufactured internally, as this definition should not be too hard to get – and to update when necessary; let us assume that the database builders just forget to mention interesting points, because they concentrate on the cost figures – as the database is generated for cost purposes. But it should never be forgotten that a cost figure without the description of the product is just worth nothing.

The subject is more understandable for products bought from other companies. People who buy products are primarily interested in the function they buy (which is of course understandable) and forget to record how it is achieved. Let us give two examples which illustrate this point:

1. A product called "electronic equipment", even if its function is given (receiver, transmitter, computer, …) does not mean anything for the cost analyst if the technology which is used is not know: technology of electronic equipment changes so quickly that this information is a major "cost driver". As a first approximation the design year can be used instead (but electronic equipments are often a mix of technologies).
2. A structure of which only the mass is known cannot be used if the material, at least, is not given. For the same structure, mass and cost can change a lot if the material changes.

Information about a product is not necessarily quantitative: sometimes qualitative information (such as the material, or the manufacturer, etc.) are important and, as we will see, can easily be used. Qualitative information can be split into two parts: factual (if it expresses a fact, such as the material) or subjective (if it expresses an opinion). Qualitative information are most often very important up to the point we generally refuse to use a formula with no qualitative information because we believe that, in our world, products – even belonging to the same product family – cannot be homogeneous enough to avoid them, except in very particular circumstances.

The conclusion of this section is very clear: try to get as much information as you can.

### 6.2.2 About the Activities

Many activities may have to be carried out on products, from development, tests, manufacturing, management, quality control, documentation, packaging, transport, maintenance, modifications up to decommissioning.

It is rather obvious that the activities which will appear in the costs have to be the same in order to make us able to compare the costs.

This may be very clear if the activities are clearly listed with the products. But, and this is of course particularly true for bought products, it may happen that costs of several activities were not added to the products costs, because (as management for instance) they appear elsewhere in the project if the project is large enough, or are "hidden" in the cost. For instance it may happen that the product you order may have to be slightly modified to fulfill your needs and that the cost of the modification is not clearly separated; it also happens that the cost of packaging is distinguished by some companies, not by other ones, etc. Hidden costs may be necessary as a provision for maintenance.

One thing the cost analyst should investigate in depth is the distinction between development costs and production costs. Many proposals, dealing with several products, succeed quite well in making it impossible to allocate to each product development and production costs, a common tactic being to use different WBSs for both: the PBS for production, the company breakdown structure for development. Both are certainly needed, but both should be submitted for development and production.

Another useful tactic, when two products have to be developed at the same time, is to split the production cost according to both products, but to add together the development costs, the rationale being that there are commonalities between the products which makes splitting the development cost impossible. There is certainly some truth about it; nevertheless if we go down one level in the product tree (at the module level for instance), some distinctions can certainly be made.

### 6.2.3 About the Over-Costs

This subject is related to the previous one, but is more general. Let us suppose we manufacture (at different times) similar products.

Even if the natures of the costs are well defined and look similar, one product may have been subject to costs that the other ones did not have. Many reasons may explain this situation: the product may have to be redone, or sent back to the factory for another test, or may have included a defective part which had to be replaced, or had to be done on overtime in order to save time, or had to be subcontracted, ...

The cost analyst should first be aware of these questions, second to be able to subtract these over-costs from the cost figures before he/she can compare them. This is not so easy and some experience is obviously necessary to carry out this activity.

### 6.2.4 Cost Definition

This is the easy part of the cost clarification.

First of all the costs should obviously be at the same economic conditions, in the same currency.

Second – and this especially true for the costs of projects – they should be expressed the same way, as constant or current ("as spent") or even "current constant". If the inflation is not negligible, the difference may be significant.

These points will be the subjects of Chapter 7.

Let us add one point which may happen. We will mention it for electronic products, but the same problem may arise in many projects. Suppose that our project (system) includes several different electronic boxes. The boxes are clearly defined; in addition we intend to do some testing with low quality components. For economic reasons management decides to buy all the electronic components together, the purpose being to make economies of scale. In the cost reporting of our company we get:

- on one hand the development and production costs of all the "naked" (which means with no component) electronic boxes;
- on the other hand the total cost of the electronic components for the whole system.

The cost analyst is interested in the cost of the "dressed" boxes (boxes with their components) which is impossible to know because the details are not available. Even if they were available, costs of boxes for this project could not be directly compared to cost of similar electronic boxes for a second project if this second project needs a far smaller number of components, which can be more expensive due to the reduced quantity. There is no easy solution to this question, unless we decide to work uniquely at the system level (which is not adequate as systems may vary a lot) or to apportion the total cost of the components on the various equipments according to their "naked" costs. This solution is obviously not satisfactory, especially if the projects include boxes with a low level of reliability and other ones with a high level. It may explain why costs of electronic boxes are often very scattered and difficult to exploit.

The same may be true for mechanical products manufactured at different times if a central procurement office is decided. This office contracts each year with materials suppliers on a given tonnage of materials and "sells" to program managers the materials they need at this price. This has two consequences:

1. Projects which consume very low volumes of materials pay a cheap price, whereas the few projects which consumes a large quantity pay about the right price.
2. Prices change every year, not due to the material cost on the market, but only due to the quantity which is negotiated each year between the procurement office and the suppliers. In such a situation we will observe that the cost of small projects will randomly fluctuate, as a function of the size of the large projects!

In such conditions when we try to correlate the cost to the equipment size, we may find that our expectations about the decrease of the specific[3] cost when the size grows does not really happen, or happens on a lesser extend than we expect, because there is a "distortion" in the price of the materials.

### 6.2.5 Expert Judgment

Data collection sometimes includes judgment of an expert. This judgment may be done about the product itself (from simple or very difficult), or the technologies it uses (from conventional to advancing the state of the art), about its cost (from cheap to too expensive), etc.

Such judgments are very interesting for the cost analyst and should not be neglected. However, judgments are opinions and are therefore subjective and, in order to properly use these judgments, one has to ask the question: How subjective are they?

The expert, in cost estimating, uses two scales:

1. The first scale refers to the factors (this is another name for the variables or cost drivers) which influence the cost, as in the examples given upwards.
2. The second scale is relative to the amount of influence each factor may have on the cost. Generally he/she cannot quantify directly this influence on the cost, but he/she may say that this factor is more important, let us say three times more important than another one.

The first thing which can be done about the first scale is to carefully establish the scale being used by the expert: When he/she says "difficult" was does that exactly mean? To this scale can drawings, descriptions, pictures, etc., be added which will be very useful and will certainly help him/her to reduce his/her level of subjectivity.

---

[3] "Specific cost" is the cost per unit of size (cost per kilogram for instance for hardware products).

Data Collection and Checking 91

There is not so much we can do about this side of the judgment if the scale has been properly established and is well known to the expert.

Let us then concentrate on the second scale: what we can do, if he/she agrees, is to directly check this subjectivity. The only way we can check it is to add some information to the process. This information may be added from the same expert or from a group of experts.

Let us start by the first one.

## Working with One Expert

The expert, quantitatively or qualitatively, knows that several factors – let us call them $f_i$ (for instance the number of components, or the surface finish) – influence the value of a variable, such as the cost. In order to "guess" the value of this variable, he/she, consciously or not, uses these factors plus a weight $w_i$ which has to be explicit. This means he/she must answer questions such as: What is the absolute weight of this factor? What are the relative weights of the other ones?

When the number of factors is greater than two, we must check the consistency of the expert's judgment, experience showing it is difficult for the expert to be consistent if more than two factors are involved.

What do we mean by "consistency"? Suppose the expert thinks that factor $f_i$ is four times more important than factor $f_j$, and that factor $f_k$ is three times more important than factor $f_j$. Answering another question, he/she thinks that factor $f_i$ is two times more important than factor $f_k$. Is this scale consistent? If the comparison is still manageable with three factors, it becomes impossible with seven: in such a case, there are 21 relative couples $(f_i, f_j)$ and their consistency is not intuitive!

### The Approach of Joseph M. Lambert
This approach assumes there is an objective scale of the factors, scale quantified by a normalized vector $\vec{w}$ of which transpose is defined by:

$$\bar{w} = \vec{w}^t = \|w_1, w_2, \ldots, w_i, \ldots, w_q\|$$

with $\bar{w} \otimes \vec{w} = 1$.

In order to find this vector, several questions are asked to the expert: he/she is asked to give a judgment on all the couples $(f_i, f_j)$. Therefore the expert must answer to $q \times (q-1)/2$ questions, a number which grows very fast with $q$. Let us call $a_{i,j}$ the answer to the question. J. M. Lambert thinks it is reasonable to limit the values of the $a_{i,j}$ to nine values, which is already large for an opinion, and proposes the scale of which structure is reproduced in Table 6.1.

These $q \times (q-1)/2$ judgments can be displayed according to a triangular table:

$$\left\| \begin{array}{cccccc} a_{1,2} & a_{1,3} & \cdots & a_{1,j} & \cdots & a_{1,q} \\ & a_{2,2} & \cdots & a_{2,j} & \cdots & a_{2,q} \\ & & \cdots & \cdots & \cdots & \cdots \\ & & & a_{i,j} & \cdots & a_{i,q} \\ & & & & \cdots & \cdots \\ & & & & & a_{q,q} \end{array} \right\|$$

**Table 6.1** The scale for the relative values.

| Relative weight | Definition |
|---|---|
| 1 | Equal importance |
| 3 | $i$ is slightly more important than $j$ |
| 5 | $i$ is more important than $j$ |
| 7 | $i$ is much more important than $j$ |
| 9 | $i$ is considerably more important than $j$ |

To this table can be associated a square matrix $\|A\|$ with the convention $a_{i,j} = 1/a_{j,i}$:

$$\|A\| = \begin{Vmatrix} 1 & a_{1,2} & \cdots & a_{1,j} & \cdots & a_{1,q} \\ a_{2,1} & 1 & \cdots & a_{2,j} & \cdots & a_{2,q} \\ \cdots & \cdots & \cdots & \cdots & \cdots & \cdots \\ a_{i,1} & a_{i,2} & \cdots & 1 & \cdots & a_{i,q} \\ \cdots & \cdots & \cdots & \cdots & \cdots & \cdots \\ a_{q,1} & a_{q,2} & \cdots & a_{q,j} & \cdots & 1 \end{Vmatrix} = \begin{Vmatrix} 1 & a_{1,2} & \cdots & a_{1,j} & \cdots & a_{1,q} \\ 1/a_{1,2} & 1 & \cdots & a_{2,j} & \cdots & a_{2,q} \\ \cdots & \cdots & \cdots & \cdots & \cdots & \cdots \\ 1/a_{1,2} & 1/a_{1,j} & \cdots & 1 & \cdots & a_{i,q} \\ \cdots & \cdots & \cdots & \cdots & \cdots & \cdots \\ 1/a_{1,q} & 1/a_{2,q} & \cdots & 1/a_{i,q} & \cdots & 1 \end{Vmatrix}$$

This matrix has an interesting property: if it is consistent (which means that $a_{2,3} = a_{1,3}/a_{1,2}$ or that the importance of factor 2 compared to factor 3 is equal to the importance of factor 1 compared to factor 3 divided by the importance of factor 1 compared to factor 2), then its eigenvalues are given by the set $[Q, 0, \ldots, 0]$. Simultaneously the eigenvector associated with the first eigenvalue is precisely equal to the vector $\vec{w}$.

In conclusion, the simplest way to check the consistency of the factors requires to compute the eigenvalues of the matrix $A$ and to check if they are close to the set $[Q, 0, \ldots, 0]$. From that J. M. Lambert built a "consistency index" which gives a figure on how consistent is the matrix.

## Working with Several Experts

Reconciliation of the judgments of several experts has been studied for quite a long time. The procedure which got some recognition is called DELPHI.

It is not our purpose to describe here this procedure which is well known. Let us just summarize the basic idea: DELPHI, as it was defined by Olaf Helmer in 1967, is the systematic use of the intuitive judgment of a group of experts in order to obtain an agreement of opinions. DELPHI works in several steps:

1. A group of experts is established.
2. A set of questions is prepared.
3. Each expert has to answer, individually, the questions.
4. A statistical summary of the answers is realized by the animator.
5. This summary is given individually to each expert: this person is then asked if he/she wants to modify his/her judgment in the view of this summary.
6. We return then to step four and five until an agreement is obtained.

Generally the process converges after a few cycles, but there is no guarantee that it will!

DELPHI has been extensively used in the middle of the last century for prospective studies.

## 6.3 Conclusion

Whatever the data you collect, you have an opinion regarding the level of its reliability; this opinion is an answer to the important question: to what extent can I rely on this data for estimating the cost of new products?

It is therefore always recommended to add to each product in the database a new variable which can be called the "level of confidence" of the data. This variable takes a value between 0 and 1: 1 means that the data is completely reliable, 0 means it is not at all.

Do not confuse it with a subjective variable: it only means that the "weight" given to this product when you extract information from the data will be less than for the other data, whereas a subjective parameter may change the relationship between the cost and the variables. A small weight only means that the influence of the data will be decreased.

# 7 Economics

## Summary

Economics normalization/denormalization comes from the fact that our measurement unit – and this is unique in the domain of science – changes in the space and in the time. The change in the time, in a given country, is generally called "inflation" although this term is not correct, but accepted. The change in space, at the same time, comes from the fact that different countries use different currencies.

As we want to compare costs originated at different moments and in different countries, taking into account the economics is something important. Neglecting to take it into account this factor can completely destroy the quality of a model.

The change of the value of a currency is a well-known phenomenon, to which we are used. Inflation indices are published since a very long time and this change is now embedded in our culture. The only thing to remember is how to handle it, and how to use economic indices.

Using different currencies is also a very common situation. However the exchange rates to which we are very well used is not adequate to compare costs, simply due to the fact that the exchange rates, at a given time, does not reflect the economic value of the currencies, because many other factors influence the values of the currencies, from the interest rates, to the expectations about the "quality" of an economy. Consequently comparing costs given in different currencies is not so obvious as it is when only one currency is used.

## 7.1 Normalization in the Same Currency

The purpose of this section is to get the cost of a finished project – or of a product, or of an activity, etc. – at specified economic conditions.

Before we start changing these economic conditions, it is important to know at which economic conditions is the cost of the project known.

### What Do We Mean by "Economic Conditions"?

Saying that the cost of something is, for instance, €300 does not mean anything, because the value of €1 changes with time: it means that with €1 today, you cannot buy exactly the same thing as with €1 one year ago. The unit of measurement changes with the time; generally speaking – but there are exceptions – the value of a currency decreases with the time.

The economic conditions of an expense simply means that the amount of the expenses is €$300_{01.04}$ with the value of the € at the date of the expense, for instance on January 15, 2004. It is a sound practice to always indicate, when you consider an expense, the "economic conditions" of this expense. This indication generally consists in writing an index with the date of the economic conditions; for instance we write € $300_{1.04}$ the index giving the month and the year of the transaction. With the rate of inflation we presently know (about 2% per year) it is useless to indicate the day in the month, except in exceptional circumstances.

The change of the value of a currency is nowadays called "inflation". The term is not correct, because the word "inflation" means in fact an increase of the quantity of money available for the economy. Historically, it has been shown that an increase of the quantity of money induces a decrease of the value of the currency (the experience of the Spanish currency after the discovery of the New World is well known) and consequently the word "inflation" became used to refer to the value of the currency.

Inflation means a *general* change in the value of goods and services, from the haircut to the apples.

There is another cause of the change in the price of goods and services which has nothing to do with inflation as it is purely due to the supply and demand. Even with no inflation, the price of copper for instance may change from one month to another month.

As both effects are difficult to distinguish, indices do not make any difference between them: they just observe that prices change.

Change in the price of a good – whatever it concerns – is known via an index: an index is a ratio, related to this good, between its value at date $a$ and at date $b$ ($b > a$). In order to manipulate easily these indexes, a reference date is often chosen for $a$ and the value of the index at that date takes the value $I_a = 100$. If $I_b = 120$, it means that between $a$ and $b$ the currency lost 20% of its value. Therefore the new value of the cost will be:

$$\text{cost}_b = \text{cost}_a \times \frac{120}{100} = \text{cost}_a \times 1.2$$

Another way of quantifying the inflation is by using an annual (or a monthly) change, given in percentage as just illustrated.

### 7.1.1 Changing the Economic Conditions of One Expense

An expense is known at given economic conditions, the one prevailing at the time it is made, for instance at date $c$.

In order to compare it with other expenses made at different times or simply to add it to them, you must requantify it at other economic conditions. The only thing you need to know to do it is the change in the value of the currency from one date – let us call it $c$ – to another one – let us call it $d$.

The simplest way to do it is to use an index which is valid for the type of commodity you are interested in; there are a lot of such indices published in magazines.

If you do not find the index related to a particular product, you may use an index valid for the whole economy, or for a certain type of industry.

In the industry this way is considered a bit rough. One prefers to create a formula which gives the amount of the various resources which were spent for producing this commodity (such as labor, energy, steel, etc.) and the percentage of each. For instance you may have:

$$\text{cost} = 60\% \text{ labor} + 30\% \text{ steel} + 10\% \text{ energy}$$

Looking to escalation tables (there are many of them, some of them available in Internet) you learn that the index for labor was 115 at date $c$ and 122 at date $d$, that the index of steel was 134 at date $c$ and 137 at date $d$, and that the index of energy was 105 at date $c$ and 106 at date $d$. Then the price at date $d$ is then given by:

$$\text{cost}_d = \text{cost}_c \times \left[ \frac{122}{115} \times 0.6 + \frac{137}{134} \times 0.3 + \frac{106}{105} \times 0.1 \right] = \text{cost}_c \times 1.044$$

This is very simple when just one expense is considered.

### 7.1.2 Changing the Economic Conditions of a Total Project Cost

*What Is the Cost of a Total Project?*

A project is a sum of expenses paid at different moments, which means measured with different units.

It is more difficult to change the economic conditions for a project as it is for one expense: a project, especially if it lasts several years, buys the resources it needs at different times. For instance some resources were bought in April 2000, other ones in September 2000, others ones in July 2001, etc. Due to inflation these costs should not been added: for instance the same resources might cost €700 in April 2000, €710 in September 2000, and €730 in July 2001.

*How Are Known the Expenses of a Project?*
Each expense is known in the economic conditions of its payment.

Two different solutions can be used to add these expenses together:

1. The first solution is, from a cost estimator's point of view, the more realistic (but also the more costly!): all the expenses are immediately converted as if they were paid at the same date; this date is known as the "reference date" (this is the point of view of the project manager, at least in Europe). This means that the cost controller, each date he/she receives an invoice, converts its amount to the value it would have been at the reference date. For instance if he/she receives an invoice of €730 in July 2001 and if the reference date is April 2000, he/she will add to the project cost €700. This solution produces a total cost often called as "total cost in reference economic conditions (REC) of April 2000".
2. The second solution does not please the cost estimator: all costs are added to the project at the value they were paid (this is the point of view of the accountant, and, as it seems to be the case, the one of the project manager in North America). The total cost is then said to be given in "current economic conditions (CEC)" or "as spent".

## How is Known the Total Cost of the Project?

**Before Completion** The total cost of the project is the sum of expenses paid, plus the sum of the forecasted expenses. We just saw how the expenses paid can be added together.

Several solutions can be used for quantifying the forecasted expenses:

- Using the REC at the beginning of the project.
- Using CEC: this assumes that an expected inflation is estimated and taken into account in the budget.
- Using R*EC: these new economic conditions means that the forecasted expenses of the project are re-estimated at the economic conditions prevailing at the date the total project cost is computed.

Suppose that the project cost is audited each year in July; in July 2001 the forecasted expenses were given in $REC_{7,01}$. In July 2002 these forecasted expenses are then given in $REC_{7,02}$. ... This means that the forecasted expenses are, every year, corrected of the inflation measured in the past year. Obviously it makes the job of the cost estimator more painful, even if it slightly simplifies the task of the project manager.

The total project cost is then computed; its value depends on the way the past expenses are recorded and the forecasted expenses presented. In order to compare the total project cost at different times during the project life, you need to know how it is computed!

**At Completion** We are back to the previous section: the total project cost is known either in REC or in CEC.

## How to go from CEC to REC?

The cost estimator must compare costs. He/she knows quite well that figures can only be compared if they are expressed in the same unit.

Practically this person needs to use REC. If different projects are known at different REC, we will see in the next section how they can be converted to the same economic conditions.

The real problem is then first to shift from CEC to REC without entering in too many details (for instance converting each expense), but without loosing too much precision.

In order to do that we need some information about how the expenses were spread in the time: we cannot adjust the same way two project costs if the first one paid 70% of its expenses during the first third of its life, whereas the second did it during the last third.

**We Know How the Expenses are Spread According to Time** What we need to know is the amount of the expenses paid by time period; this period can be the month, the trimester or, for long projects, the year. Inside a company this information is not difficult to get. Suppose we get the information given in Table 7.1.

We can:

- either convert the expenses of each period to any REC, using an escalation table per period,
- or, if the inflation rate per period is similar, assume that all the expenses occurred at the barycenter – or center of gravity – of this distribution, which is, in this case, period 5.05.

**We Do Not Know How the Expenses Are Spread According to Time** The only information we have is that the project lasted for 10 periods of time and that its total cost in CEC amounts to 28 500.

Table 7.1 Example of the expenses paid per year.

| Period | Expense |
| --- | --- |
| 1 | 1 000 |
| 2 | 2 000 |
| 3 | 4 000 |
| 4 | 5 000 |
| 5 | 4 000 |
| 6 | 6 000 |
| 7 | 3 000 |
| 8 | 2 000 |
| 9 | 1 000 |
| 10 | 500 |
| Total | 28 500 |

We need then to modelize the expenses over this period of time. One of the easiest ways to do it is to use the BETA distribution (see Chapter 3 of Volume 2) which is often used because, with its two parameters $\alpha$ and $\beta$, it can fit a lot of real distributions. This distribution is given by the formula:

$$B(x, \alpha, \beta) = \frac{\Gamma(\alpha + \beta)}{\Gamma(\alpha)\Gamma(\beta)} \times (x)^{\alpha-1} \times (1 - x)^{\beta-1}$$

Let us assume that, from your experience (you need some experience!) you select $\alpha = 2$ and $\beta = 3$. The distribution is then given by the formula:

$$B(x, 2, 3) = 12 \times x \times (1 - x)^2$$

which is represented in Figure 7.1.

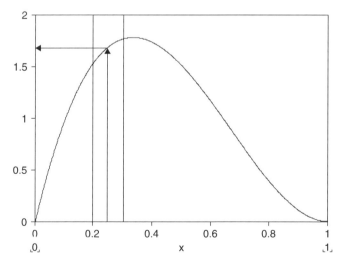

Figure 7.1 Modelization of the distribution of the expenses.

As we considered that the project lasted 10 periods of time, the abscissa is divided 10 equal parts of which the middles are given by 0.05, 0.15, etc. To each middle point corresponds a value of the distribution: for instance for $x = 0.25$, then $B(0.25, 2, 3) = 1.688$. These values are given by Table 7.2 Note that the total is not exactly equal to 10.

**Table 7.2** Values of the BETA function for the middle of the intervals.

| $x$ | BETA$(x, 2, 3)$ |
|---|---|
| 0.05 | 0.542 |
| 0.15 | 1.301 |
| 0.25 | 1.688 |
| 0.35 | 1.775 |
| 0.45 | 1.634 |
| 0.55 | 1.337 |
| 0.65 | 0.956 |
| 0.75 | 0.563 |
| 0.85 | 0.230 |
| 0.95 | 0.029 |
| Total | 10.055 |

We now assume that the expenses during a period is equal to the value of the BETA function, multiplied by 0.1 (which is the size of the period on the graph) and by 28 500 in order to retrieve the total cost. As the total of the middles values was not equal to 10, these values must be slightly corrected.

Eventually the distribution of the expenses is given in Table 7.3.

**Table 7.3** Distribution of the project costs.

| $x$ | BETA$(x, 2, 3)$ | correction | Distribution of the costs |
|---|---|---|---|
| 0.05 | 0.0542 | 0.0539 | 1 536 |
| 0.15 | 0.1301 | 0.1294 | 3 688 |
| 0.25 | 0.1688 | 0.1679 | 4 784 |
| 0.35 | 0.1775 | 0.1765 | 5 031 |
| 0.45 | 0.1634 | 0.1625 | 4 631 |
| 0.55 | 0.1337 | 0.1330 | 3 790 |
| 0.65 | 0.0956 | 0.0951 | 2 710 |
| 0.75 | 0.0563 | 0.0560 | 1 596 |
| 0.85 | 0.023 | 0.0229 | 652 |
| 0.95 | 0.0029 | 0.0029 | 82 |
| Total | 1.0055 | 1.0000 | 28 500 |

Now that the distribution of the cost is known (modelized) the same procedure as indicated upwards can be used in order to establish the cost of the project in REC (the barycenter of this distribution is equal to 4.48).

## Adjusting This Total Cost to Other Economic and Technical Conditions

We saw that a global index is frequently used to change the economic conditions of an expense, and consequently of a project.

## The Petro-Chemical Industry

The US petro-chemical industry publishes several indices (see [18], p. 212):

- ENR (Engineering News-Record) construction cost index, published every week in the magazine Engineering News-Record,
- M&S (Marshall and Swift) equipment cost index, published in the magazine Chemical Engineering,
- Nelson refinery construction cost indexes, published in The Oil and Gas Journal.

The reader will consult these magazines or the Ref. [18] to get a definition of these indexes. This book presents the Nelson index (one of the mostly used) from 1950 up to 1999.

## 7.2 Between Two Countries

The problem to be solved here is the following one: we know the cost, let us call it $C_X$, of an object $O_X$ (which can be a product or a whole plant) made in country $X$ (and therefore $C_X$ is known in the currency of this country and at economic conditions $E_X$) and the cost, let us call it $C_Y$, of an object $O_Y$ (which may be or not a similar object[1]) made in country $Y$ (and therefore $C_Y$ is known in the currency of this country and at economic conditions $E_Y$). We want to compare these costs.

Pay attention to the fact that country $Y$ does not want to buy object $O_X$; in such a case the cost estimator should use the results of the previous paragraph in order to adjust the economic conditions to the ones prevailing today in country $X$ and then multiply it by the today exchange rate. This will give him/her the acquisition cost.

No! What we want is to compute the cost of object $O_X$ as if it were made in country $Y$. This type of problem arises in two slightly different situations:

- Country $Y$ wants to produce an object very similar or even – in order not to make the problem too complex – identical to $O_X$. An estimator is in charge of providing an estimate in the currency $Y$; but this estimator never had to deal with this type of product before (and therefore a specific[2] model was never built for it) and does not have enough information about it to be able to use a general[3] model. Fortunately enough he/she knows the cost $C_X$.
- You, as an estimator, collected the cost of several products belonging to the same product family; these products were made in different countries, each one with its own currency. You would like to create a specific model from these cost data. This requires that the cost unit must be the same.

## A Conventional Approach

This approach is largely used by companies who make plants all over the world, especially chemical plants or petrol refineries. These companies have urgent needs

---

[1] The question "Is it realistic to compare them from a technical point of view?" is irrelevant here.
[2] "Specific models" are defined in Chapter 1 of Volume 2.
[3] "General models" are defined in Chapter 11.

to forecast costs in any country during what we called the decision phase: they know what the cost would be for the required capacity in one country, but how much will it cost in another country?

The interested reader will find in Ref. [18] the approach used by these companies.

## A General Approach

The general approach does not make reference to any product. It aims at substituting to the exchange rate – which cannot be used because it changes everyday due to the demand and supply for the currencies – some "conversion factor" which would not be subject to these variations and would therefore convey the economic value of the currencies. This rate will be called here the "economic conversion factor" or "ECF" between the currencies.

The idea to find out this factor is rather basic: it starts from the hypothesis that, if the transfer of goods and services on one hand, and of money on the other hand, are free between two countries, then this factor could not be – on the long run, because the supply and demand may modify the exchange rate on the short run – different from the exchange rate. Otherwise if the exchange rate is permanently under or over quoted, then a flux of goods and services will re-establish this exchange rate at its "normal" value. Let us remind it: it assumes the freedom of exchanges.

Then the question is: what do we mean by the long run? As economies do not adapt to each other very quickly (and therefore the adjustment of the exchange rate cannot be done in a couple of days), the long run should mean several years. In order to take into account the fluctuations of the exchange rate due to the supply and demand – fluctuations of which the damping time is not well known – economists suggest to consider at least a period of 10 years.

This ECF has no reason to remain stable (otherwise a simple average of the exchange rate would solve the problem). This factor compares the real value of the monetary units in two countries: how much would cost you the same amount of goods and services in both currencies? An economy is not something completely stable: economy can very well deteriorate in one country whereas it does not in the other country. The ECF is a function of the date at which it is computed.

A second point to take into account is the fact that the inflation rate in both countries is generally different. In economies fast to react, the difference of inflations should automatically adjust the exchange rates.

Therefore the solution of the problem (finding a "dynamic ECF") needs three types of information: both inflation rates over a long period, plus the exchange rate over the same period. As inflations are generally quantifies by indices[4], let us call:

- $I_{X,\text{date}}$ the inflation indices in country $X$ (column 1 of Table 7.4 – illustrated for France).
- $I_{Y,\text{date}}$ the inflation indices in country $Y$ (column 2 of Table 7.4 – illustrated for the USA).
- $E_{\text{date}}$ the exchange rate (column 3 of Table 7.4: so many French francs for US$1).

---

[4] The indices, because we are comparing the whole economies, should refer to the global level of prices and not to a particular industrial sector.

**Table 7.4** Computation of the ECF (sample).

| 1966 | 1 | 2 | 3 | 4 | 5 |
|---|---|---|---|---|---|
| January | 327 | 385 | 4.902 | 4.661 | 1.000 |
| February | 327 | 385 | 4.902 | 4.661 | 1.000 |
| March | 328 | 386 | 4.901 | 4.663 | 1.000 |
| April | 329 | 387 | 4.901 | 4.665 | 1.001 |
| May | 330 | 389 | 4.901 | 4.655 | 0.999 |
| June | 330 | 390 | 4.901 | 4.643 | 0.996 |
| July | 331 | 391 | 4.901 | 4.645 | 0.997 |
| August | 332 | 393 | 4.903 | 4.636 | 0.995 |
| September | 332 | 393 | 4.923 | 4.636 | 0.995 |
| October | 333 | 395 | 4.939 | 4.626 | 0.993 |
| November | 334 | 396 | 4.943 | 4.628 | 0.993 |
| December | 335 | 397 | 4.951 | 4.630 | 0.993 |

The procedure then takes several steps:

- a reference date is selected, let us call it "0",
- the relative inflation between both countries is computed by (column 5 of Table 7.4):

$$RI_{date} = \frac{I_{X,date}}{I_{X,0}} \bigg/ \frac{I_{Y,date}}{I_{Y,0}}$$

- then we assume that the relationship between ECF and RI should be given by $ECF = A \times RI^B$, formula which linear in log–log. The reason for choosing such a formula instead of a pure linear formula is not very clear: let us say it is the weight of the tradition, as cost estimators know that such a "multiplicative formula" generally gives more satisfactory results. How can then be found $A$ and $B$? Very simply by making a regression (the word is defined in Volume 2) between the exchange rate and the relative inflation: the result will give the ECF.

Making such an ordinary least square (OLS) regression on the logs on 293 months (much more than the 10 years recommended by the economists!) gives

$$A = 4.663$$
$$B = 0.988$$

The value of B "proves" that the hypothesis that the change of the exchange rate is strongly – on the long run – correlated to the relative inflation is correct.

- from this equation the ECF is computed for every month (column 5 of Table 7.4).

Reproducing all the values is not the purpose of this section. Just for illustration, Figure 7.2 displays the relative inflation, the exchange rate and the ECF (thick curve) between France and the USA. It clearly shows that:

- the short term variation of the exchange rate are deleted (the linear regression is a "low pass filter"),
- the ECF is strongly correlated to the relative inflation.

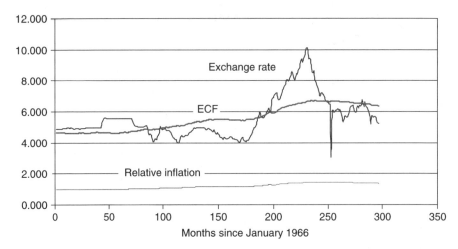

**Figure 7.2** ECF between France and the USA.

It is interesting to observe that the ECF as such computed is very close to the exchange rate which was decided at the date the Euro was created (6.6685 FRF/€).

## 7.3 Conclusion

Given different costs in different countries can now be "normalized" to the same currency, which is going to be our cost unit.

This first normalization is the first one to take into account, just in order to use the same unit for the costs we handle.

# 8 The Cost Improvement Curves

## Summary

Cost improvement curves refer to the fact that, when you do two times the same thing, you need less time to do it the second time than for the first. The difference of time may be large (for instance 80%). As we want to compare products, we have to be sure that we compare similar cost figures: comparing a product of production rank 1 with a product of production rank 10 would not be realistic. Taking into account the change in cost when producing several identical pieces in the normalization process is therefore important.

The cost improvement has been modelized long time ago (practically in the middle of the 20th century). At that time two different – but looking similar – models were proposed by:

- by Wright, based on the average production cost;
- by Crawford, based on the unit production curve.

This chapter introduces these models and show they produced similar results, but based on different parameter values. The models must be clearly distinguished because they look so similar that confusion between them is frequent and leads to mistakes.

It then shows that in practice these models must be modified as soon as the production quantity becomes large, "large" meaning as few as 10 for complex items (then the cost improvement curve refers mainly to "learning"), or as large 10 000 for very simple items (then the cost improvement curve refers mainly to "reflex acquisition"). This explains why simple models found in the literature should not be applied without any reference to actual cost data for comparable products.

More complex models have been proposed by different authors. This chapter mentions them without entering into too many details.

## 8.1 Introduction

*You can skip this chapter if you are producing "objects" one at a time.* This is the case for instance if you build, from time to time, petro-chemical plants. There is also some "learning" in this process, but it is of a different type of what we are dealing with here: we call then it "experience" and it is subject to different rules.

In this chapter we consider the repetitive and continuous production of similar products, one after the other. The produced quantity might be small or large, but this chapter does not deal with mass production: the rules we will observe or establish here do not apply to this case. We will see why in Chapter 9.

Production is achieved by **activities**. The first analysis of the phenomenon (in 1936) we are interested in here was, for this reason, made on activities: this is why it is still sometimes called "learning". But it can as well be applied to full production cost: in this case, the terms "cost improvement" is generally used.

The concept of learning or cost improvement must be studied for at least five reasons.

First of all, learning – or cost improvement – is an important industrial phenomenon: it may reduce the time – and consequently the cost – for performing an activity by a large factor, as we will see it in the examples. Therefore it has to be known and applied by the cost estimator.

The second reason is related to the process for normalization/denormalization mentioned in the previous chapter:

- It may happen you want to compare (for instance to build a specific model as we will do it in Volume 2) two products. The first one was the second identical one to be produced; for the second one the only thing that you know is that it was produced during a batch which contained five identical products, the first one of this batch being preceded by another batch of 10 similar products. Due to the cost improvement process we will deal in this chapter, these products cannot be directly comparable: what you would like to have, in order to make this comparison useful, is the cost of both products produced at the same rank (generally we choose the first production rank). This chapter will allow you to do it, which means to "*normalize*" the costs of both products in order to know what they would have costed if they were the first ones to be manufactured.
- From the comparison you make about these two products, you are now able to estimate the cost of a third – different – products (we will see how in the next chapters). The cost you will compute will be the manufacturing cost of the first product of this species to be produced; now the decision-maker would like to know what is going to be the average production cost if you produce 100 of them. We enter now in the process of "*denormalization*".

The third reason is specific to this book: in order to make it usable by the cost estimator, we need to give him/her a tool to work with. This tool will be what we will call in the next chapters a "parametric model" of the phenomenon: it will therefore introduce the reader to the subject of this book and help him understand why we call it this way, how it is built and how it is used.

The fourth reason is that learning curves are now part of the culture of any manager: any decision-maker – and naturally any procurement officer – knows about cost improvement curves. They consequently not only expect to see them appear, but will also challenge the figures used by the cost estimator: in many negotiations, a large percentage of the time may be devoted to arguing about this concept and how it was applied.

The fifth reason is that improvement curves are built in all the commercial models. In our opinion, the way they are used in these models is not always "perfect" and the user of these models should know how much he/she can rely on them.

This chapter is then important for practical and theoretical reasons: it will offer a useful tool to the cost analyst and prepare the following chapters.

The same concept of learning applies as well to activities – measured in time – as to **costs** – measured in €, in £, in $, in ¥, or in any other currency, as long as the currency is "constant" (see Chapter 7). Originally, when discovered by Wright, it was applied to activities. Later on it was found, also by Wright, that it can also be applied to products costs for several reasons:

- Activities are the base of products manufacturing.
- Using machine tools requires two operations: preparing the machine and using it. The cost of preparation is allocated to all the products manufactured in the same "run": the larger the quantity, the smaller the allocated cost per product.
- The materials are the second important part in products costs. It is well known that, if you buy 1000 kg of material, you will pay less per kilogram than if you buy just 10 kg: this is not "learning" (the decrease is known as Young's law) but contributes to the decrease in products costs.
- The same is true for all the other components of products costs; for standard elements (such as the energy) you cannot expect cost changes, but this is not the case for specific components of which costs must follow some learning curve.

There is however an important difference between activities and costs: the so-called "learning curves" apply well to individual activities as well as to their sum (we will speak about "unit time" for making one activity, and "total time" for making several times the same activity). These curves apply also well to the cumulative cost of several identical products, but not (except for their relevant activities) to individual products: for instance the material cost, when you manufacture the first or the thousandth product, may cost the same; but the unit material cost for manufacturing 1000 products will cost less per product than what it would cost if you manufacture one product only (unless if you buy large quantities of material, because the same material is used for other projects).

Nevertheless, we will use indifferently the word "activity" or "product" in this chapter. It will be let to the user to determine what concept may be used or not.

A last observation: Wright himself found that the learning rates for products costs were always higher (there is less learning) than for activities; he found, for products costs, learning rates between 0.83 and 0.87 instead of 0.80, for activities. Therefore, never apply learning rates observed on activities to products costs and inversely, without checking the rates.

Let us now introduce the subject.

You know quite well that, if you repeat several times the same activity in the same environment (think about boring holes in the walls of your home, about driving from home to your work place in a new location, about filling the same administrative form for several persons, …), it takes less and less time to perform it. This is this time reduction observed when we perform several times – one after the other with no interruption – the same activity that we call "learning".

The same concept applies to any human activity, from walking to driving a car, and to all production activities. It is therefore very important for the cost estimator.

This chapter describes how this phenomenon can be taken into account for cost estimating.

## General Information About "Model" Building

How can the cost estimator apply a concept? As we do in all engineering activities, we need to give him/her a quantitative "recipe" which takes nowadays[1] the form of an **algorithm**.[2] How do we build such an algorithm? As any engineer knows it, **it involves three to four steps:**

1. *Getting the facts*: Here it means collecting the time it takes to perform several times the same activity.
2. *Finding the general trend(s)*: This is an important concept when we try to apply engineering thinking to human activities which does not really exist in other engineering studies: if you make several times the same physical or chemical experiment in the same conditions in order to measure some result, you will observe that this result fluctuates slightly from one test to the other one. This comes from the fact that the measurement process is not perfect: when you read the result of an experiment on an instrument, you make small errors; these errors have nothing to do with the experiment itself (except if you study the motion of individual particles …): we believe that nature follows some rules. This is not the case when you study human activities: even in the absence of learning, a human activity never takes exactly the same time to perform: we are not robots … . This is a very important feature of human activities, which, by the way, sets a natural limit to our capacity to make cost forecasts; we will return several times to this idea in this book. What we want to insist on here is that the cost analyst, before he/she can develop an algorithm, must interpret the data and "extract" from them the general trends, by removing the – not negligible – random fluctuations from the measurements. This is not always an easy step because if he/she removes too many "fluctuations", he/she may overlook some interesting phenomenon: we will see an example in the next section.
3. *Introducing, maybe, some parameter(s)*: If a unique trend is found in the data, there is no need for a parameter. If several trends are perceived – and if the cost analyst considers they should be preserved – then parameters have to be introduced. This is our first contact with the subject: here a parameter is just a variable which takes such or such value that the cost analyst *uses to discriminate between different trends* he/she discovers. This will be clearer on the examples.
4. *Establishing the algorithm*: Generally speaking, it involves writing mathematical formulae. The formula can be different for each value of the parameter(s), but it is considered aesthetically important – and, we must admit, easier to use – to have a general formula which can take different forms depending

---

[1] Not so long ago the "recipe" was a chart or a set of figures grouped into tables. You can still find such tables in some books. As now everyone has his/her own computer, it is much easier to use mathematical formulae than tables or even charts.

[2] You may not be familiar with this word which will be largely used in this book. An algorithm (the word comes from the Arab mathematician Al-Khwarizmi who lived during the 9th century) is a set of rules that you must follow to solve a problem. Nowadays the rules are most often mathematical relationships: an example is given by the solution of a second degree equation: compute the value of the equation determinant; if it is positive, apply such formulae.

on the value of the parameter(s). There are several ways to find out the algorithm:
- *Trial and error*: this is the most common way! The speed of the process depends on the cost analyst genius …
- *Curve fitting*: there are some mathematical packages which elaborate formulae from a set of data points. The result is generally complex …
- *Statistical analysis*: statistics allow – generally by regression analysis (this procedure will be discussed in details in Volume 2) – to find simple formulae to represent trends. Residuals must be carefully investigated by the cost analyst.

The one we will favor at this stage is the first one. Other ones will be used later on.

Once the four steps are accomplished, the cost analyst gives the cost estimator the algorithm, plus some instructions on how to use it …

Let us apply this introduction to our subject of learning.[3]

## 8.1.1 Getting the Facts?

Here is an example (source: Ref. [7], p. 565) of the time needed to perform several (thousands!) times the same activity.

This example gives the time observed ("time per cycle") – which is really an average time per day – for a small mechanism assembly operation; in abscissa, is the number of hours worked (Figure 8.1). Generally, such a curve is given in setting in abscissa the number of cycles performed; drawing such a graph implies to

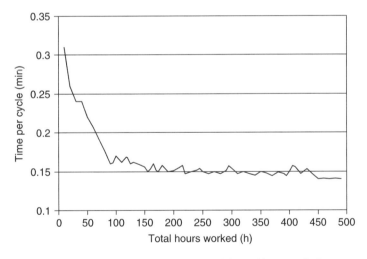

**Figure 8.1** Time per cycle as a function of the total hours worked.

---

[3] Old books about the subject are filled with tables or graphs allowing an easy computation of the algorithms. Nowadays, with the availability of powerful hand calculators, there is absolutely no need for such tables and graphs. All the formulae given here can be easily computed with a basic scientific (because log or exponentiation are used) calculator.

compute this number from the hours worked and the time per cycle. The result is given in Figure 8.2.

What is obvious on this Figure 8.2 is that the time per cycle decreases, on an average, with time.

This is the decrease in the time per cycle which is called "learning".

### 8.1.2  Finding the General Trend(s)

Figure 8.2 clearly shows two things:

1. First of all we observe fluctuations in the time devoted to each cycle. This is, as we mentioned it, typical of human activities. The cost analyst must eliminate these fluctuations without deleting the phenomenon he/she is looking for: it is not really difficult when we realize the fluctuations we want to remove are "short-term" fluctuations,[4] whereas the interesting phenomenon is "long term";[5] it should not be too difficult to automatize short-term fluctuations removing,[6] but, in the present situations, we will use our intuition. Before abandoning the subject, it is interesting to note the magnitude of these short-term fluctuations: they are of about ±0.01 min or about 6%; in the present situation, it will be the limit of any

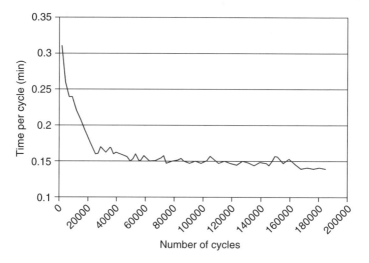

**Figure 8.2**   Time per cycle as a function of the number of cycles.

---

[4] The fluctuations we observe here are fluctuations on the average time per day. The "very short-term" fluctuations – from one cycle to the following one – are already eliminated by this averaging process.
[5] We will call "long-term" effects what remains when the "short-term" fluctuations are removed. Instead of "long term", we could have said "general trend".
[6] The process is called "filtering": the purpose is to "filter" (remove) the short-term fluctuations in order to make appear the long term ones. In doing that the cost analyst must decide on the "term": choosing a too long term may eliminates interesting things, choosing a too short-term lets uninteresting fluctuations appear. In this example, the eye remains a powerful tool to help make a decision.

forecast on the average (per day) time per cycle (the limit on the real time per cycle would probably be much larger).
2. Second it is very dubious that just one trend appears. Of course we have some experience about the subject, but the evidence of two or three trends may not be obvious to the person who analyzed this kind of thing for the first time. What must attract his/her attention is the change of slope at about 22 000 cycles: before this number of cycles, we note a sharp decrease in the time per cycle and then suddenly after 22 000 cycles the rate becomes rather small (it becomes even smaller after 50 000 cycles, but this new change in the slope is not as visible). The cost analyst must always pay attention to this kind of thing!

What we can say at this stage is that there are at least two or, maybe, three "long term" trends: from about 1 to 22 000 cycles, from about 22 000 to 50 000 cycles, and beyond 50 000 cycles. For simplicity, we will keep here only two trends, guessing the second and third trends can be assimilated.

### 8.1.3 Introducing Parameter(s)

We want to use (this is our objective: find out a simple recipe) just one algorithm only to "modelize" all trends: as the trends do not really have the same slope, we will use a parameter to characterize them.

Another thing which should enter in the algorithm is the starting point, which means the time spent for making the first cycle: if we were looking at another operation (such as making a hole, or filling a form), we would probably have observed a curve parallel to this curve because, due to the nature of the operation, it could have started higher or lower than the studied time per cycle.

This starting point is not of the same nature as the parameter we introduce in the first sentence of this section. The name "variable" should be more appropriate. Here is the rule we will follow through this chapter:

- A **parameter** is a value that changes the behavior of the "model", whatever is the operation been studied: here we will see that the parameter changes the rate of the decrease in time per cycle. It is quite possible that the rate will be the same for different operations.
- A **variable** is a value which is specific to the operation been studied. It depends on the nature of the operation only. The starting point we just saw is a very common variable, up to the point it receives a special name everyone recognizes: it is called the "first piece cost" – the acronym "FPC" which is used quite often and we will follow this usage.

  Quite often this FPC is called "theoretical first piece cost" or "TFPC". But this acronym has two different meanings in the cost analyst vocabulary:
  – This FPC can be called "theoretical" because it does not really exist. There is nothing really equal to a FPC because when a new operation is started for the first time, the time to do it fluctuates a lot and it would be difficult to say: "here is the real FPC"; the cost of the second operation could be higher than the practical first one. Therefore what is called here the FPC is the cost which can be computed if we extrapolate backwards the trend until we get a (mathematical) value for the cycle no 1.

- In the conceptual models, it may happen that a theoretical FPC is computed; this cost is afterwards modified to take into account other parameters. The FPC is then just a temporary result.

So we have now a *variable specific to the operation* and *a parameter which*, we hope (at this stage!), *correctly "modelize" the change* and could be applied – maybe with a different value from one operation to the other one – to any situation.

Although we previously separate steps 3 and 4, it is clear they go together, hand in hand: the value of the parameter depends, as we will see it, on the algorithm and the algorithm depends on the type of the parameter.

### 8.1.4 Establishing the Algorithm

Now we have to utilize the language of mathematics. We will use the following notations:

- $t_i$ the unit manufacturing cost of product of rank $i$, or the time to perform the activity of rank $i$ (the first product manufactured or activity performed will be said "rank 1", the second one "rank 2", etc.).
- $T_n = \sum_{i=1}^{n} t_i$ the total time for producing the first $n$ products or performing the first $n$ activities.
- $\bar{t}_n = T_n/n$ the average time[7] for producing the first $n$ products.

If we deal with a set of products from rank $m$ to rank $n$ included (this is called a "lot" of which size is $Q = n - m + 1$):

- $T_{m,n} = T_Q = \sum_{i=m}^{n} t_i$ the total time for producing this lot (with, of course $T_{1,n} = T_n$).
- $\bar{t}_{m,n} = T_{m,n}/(n - m + 1) = T_Q/Q = \bar{t}_Q$ the average production time for the products in this lot.

Several algorithms have been proposed by different persons in order to modelize the trends we saw; these algorithms are generally called "laws", even if in this book the word "model" will be preferred. We will here introduce with some details the two basic – or "pure" – laws and then mention how these laws, which are now part of the cost analyst "culture", have to be modified for practical use:

- The first algorithm is called "Wright's law". It modelizes the average times $\bar{t}_n$.
- The second one is called "Crawford's law". It modelizes the unit times $t_i$.

Both laws are practically used by all cost estimators. From our experience, the first one seems to be used more often in the USA, the second one in Europe; but it is just an opinion and it does not really matter ... if you know which one is used. We will return to this point once both laws are described.

---

[7] A small bar over the name of a variable generally means (in all statistical manuals) an average. In order to allow you reading these manuals, we will use the same notations.

# 8.2 The Continuous Production: the Simple or "Pure" Models

## 8.2.1 Wright's Law (Deals with Average Time)

The discovery of the learning curve is often attributed to Wright. It is said[8] that, during the year 1920, not only Wright observed the phenomenon, but also measured it, modelized it and discovered that what he called the "learning rate" remained constant. His results were published (*Journal of Aeronautical Science*) in 1936.

Wright defined the **learning rate** – that we will designate by $\lambda_W$ ($\lambda$ reminds us we are speaking about learning, W reminds that this is Wright's rate) – as the factor by which the average time needed to produce a given quantity Q of identical objects (in his studies aircraft cells) must be multiplied to get the average time needed to produce a double quantity 2Q of the same objects. In mathematical terms, he wrote:

$$\bar{t}_{2Q} = \lambda_W \times \bar{t}_Q \tag{8.1}$$

The value he found for $\lambda_W$ was close to 0.8 (note that this value must be lower than 1; otherwise no learning would be observed). Let us illustrate with figures: it means that if you need an average manufacturing time of 15 min to produce 10 objects (the total manufacturing time is then 150 min), you will need an average manufacturing time of 12 min to produce 20 objects (the total manufacturing time will be 240 min). This is the change in the average manufacturing time which he called "learning".

Generally speaking, the most frequent use of the law is to compute the time which is needed to manufacture a quantity *n* of the objects, starting at rank 1. The law then takes the following form[9] which is the form under which it is generally known (note that the term W, called the **learning slope**, is a negative number, because $\lambda_W$ is lower than 1):

Wright's law on the **average unit time**:

$$\bar{t}_n = t_1 \times n^W \quad \text{with } W = \frac{\log \lambda_W}{\log 2} \tag{8.2}$$

$t_1$ is the variable we called FPC (which depends on the type of the manufacturing operation), $\lambda_W$ being the parameter.

Let us check if this "law" gives the same result as Eq. (8.1). The average time to make the first *n* operations is given, from Eq. (8.2), by:

$$\bar{t}_n = t_1 \times n^W$$

whereas the time to make the first 2*n* operations is given by:

$$\bar{t}_{2n} = t_1 \times (2n)^W$$

---

[8] *The learning curve deskbook. A reference guide to theory, calculations, and applications.* Charles J. Teplitz. Quorum Books. 1991.
[9] The log function can be the natural log (base e) or the decimal log (base 10). It does not matter here because we use a quotient of two logs.

The ratio between these two average times can be written as:

$$\frac{\bar{t}_{2n}}{\bar{t}_n} = 2^W$$

and $2^W = \lambda_W$ because:

$$\log\left(2^{\frac{\log \lambda_W}{\log 2}}\right) = \frac{\log \lambda_W}{\log 2} \times \log 2 = \log \lambda_W \quad (8.3)$$

Therefore formulae (8.1) and (8.2) are two sides of the same coin. If $Q$ operations do not start from the first one, computation is a bit more difficult, but the result is the same.

Let us illustrate Wright's law on an example. Suppose $t_1 = 3\,\text{min}$ and $\lambda_W = 0.8$; the average time to make the first 10 operations is given by (8.2): $3 \times 10(\log 0.8/\log 2) = 1.43\,\text{min}$. The reader immediately understands from this example that learning is a very important subject in cost estimating: in this example, the average time for the first 10 operations is divided by more than 2! Figure 8.3 illustrates Wright's law for two values of $\lambda_W$: 0.8 (full line curve) and 0.9 (dotted line curve); for this figure we chose $t_1 = 1$. There is no need to represent the law for $\lambda_W = 1$: it would be a straight horizontal line for an average time of 1.

Note that the decreases in time are rather sharp for a small number of operations, but become moderate when this number increases.

The trends are easier to perceive when one uses the log–log scales, because the law is then represented by a straight line, as Figure 8.4 illustrates.

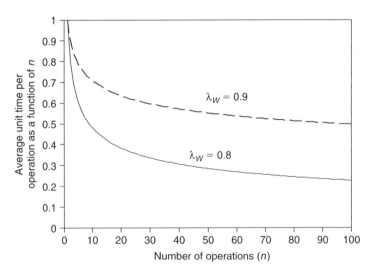

**Figure 8.3** Wright's law (average time for $n$ operations) for $\lambda_W = 0.8$ and 0.9 (linear scales).

The Cost Improvement Curves

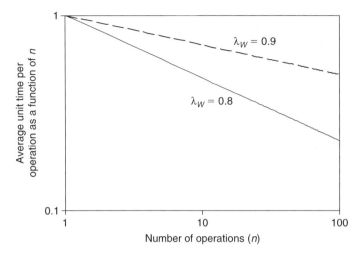

**Figure 8.4** Wright's law (average time for $n$ operations) for $\lambda_W = 0.8$ and $0.9$ (log–log scales).

## What About the Total Cost?

The total cost is easy to compute here: as we know the average unit time for $n$ operations, the total time to perform the $n$ operations will simply be the product:

$$T_n = n \times t_1 \times n^{\frac{\log \lambda_W}{\log 2}} = t \times n^{1+\frac{\log \lambda_W}{\log 2}} \tag{8.4}$$

It is represented on Figure 8.5 (with $t_1 = 1$): of course it grows with $n$, but at a slower pace than $n$; the reader will notice that, when $\lambda_W = 0.9$, then the total cost

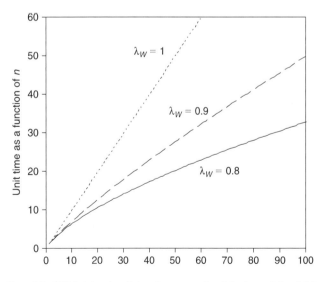

**Figure 8.5** Wright's law (total time for $n$ operations) for $\lambda_W = 0.8$ and $0.9$.

for 100 operations is exactly half what it would be if there was no learning: it illustrates how important may the learning be!

On this figure we added the straight line when there is no learning ($\lambda_W = 1$). Of course the total cost becomes then a straight line with a slope of 1 (due to the fact that we chose a $t_1 = 1$) when there is no learning.

## What About the Unit Time?

As we said and repeated it, Wright's law deals with **average unit** time for a number of operations. Quite often however cost estimators have to answer the question: What will be the unit (not average) time for the $p$th operations? We therefore have to compute the unit time from the average time given by the law.

This unit cost is not so easy to compute for Wright's law. The only thing we can do is to compute the total cost for $p$ operations and subtract from it the total cost for $p - 1$ operations:

$$t_p = t_1 \times p^{1+\frac{\log \lambda_W}{\log 2}} - t_1 \times (p-1)^{1+\frac{\log \lambda_W}{\log 2}} \qquad (8.5)$$

This unit time decreases rather quickly (Figure 8.6): even for $\lambda_W = 0.9$ (which is not abnormal), one clearly sees that the unit cost, compared to the FPC, is divided by about 2 for $p = 40$!

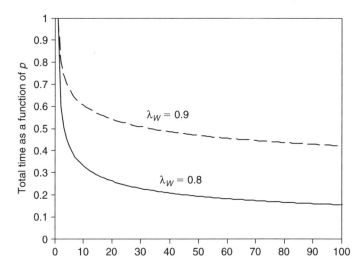

**Figure 8.6** Wright's law (unit time for $p$ operations) for $\lambda_W = 0.8$ and $0.9$.

## What About a Production Lot?

The total production cost of a lot starting at rank $m$ and finishing at rank $n$ is easily computed:

$$T_{m,n} = T_n - T_{m-1}$$

## 8.2.2 Crawford's Law (Deals with True Unit Time)

Crawford's law is more recent than Wright's law: studies carried out after World War II (Ref. [29], confirmed by a study of Stanford University (1949)) showed that the production of aircrafts followed the type of law discovered by Wright but that this law should be applied to true unit time and not to average unit time. Crawford's law seems to be more frequently used nowadays, especially in Europe, but Wright's law is still sometimes used: the cost estimator should know both.[10]

Crawford's law has the same mathematical expression as Wright's law but deals now with true unit time. Therefore we write:

Crawford's law on **true unit times**

$$t_n = t_1 \times n^C \quad \text{with } C = \frac{\log \lambda_C}{\log 2} \tag{8.6}$$

where $C$ as an exponent or an index reminds the reader that we are dealing now with Crawford's law. Note that $n$ does not refer now to the number of operations (which is a cardinal number) but to the rank of operation (which is an ordinal number: 1st, 2nd, etc.). $t_1$ still refers to the FPC.

The shape is exactly the same as Wright's law, but, let us insist on that, deals with true unit time. It is reproduced on Figure 8.7 for the sake of convenience.

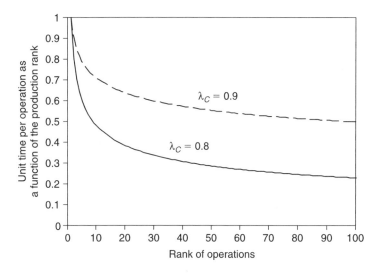

**Figure 8.7** Crawford's law unit time for $\lambda_C = 0.8$ and $0.9$.

---

[10] Some authors use the generic term "Wright's laws" for several learning models (the explanation is that Wright's law came first and that, consequently, the name of Wright became attached to the learning concept). We prefer to clearly distinguish them.

Crawford's law has the same meaning as Wright's law, transposed on unit time: each time the production rank $i$ is doubled, the unit time is multiplied by the learning rate $\lambda_C$ (which must be lower than 1):

$$t_{2i} = \lambda_C \times t_i \tag{8.7}$$

## What About the Total Cost?

The total cost up to rank $n$ (starting at rank 1) is given by:

$$T_n = \sum_{i=1}^{n} t_i \tag{8.8}$$

It would be very cumbersome to compute this value, as soon as $n$ is large. Fortunately the mathematician Simpson developed a formula giving an excellent approximation (we use the symbol, very frequent in statistic – as we will see it –, to represent an estimate), valid even when $n$ is very small (for $n = 1$ or 2, the ratio between the estimated value and the true value is about $10^{-3}$, which is largely sufficient for cost estimating):

$$\hat{T}_n = t_1 \left\{ \frac{1}{1+C} [(n+0.5)^{1+C} - 0.5^{1+C}] - \frac{C(C-1)}{24} \right\} \tag{8.9}$$

Some authors (for instance Charles J. Teplitz. *The learning curve deskbook* (1991), p. 41) recommend another formula:

$$\hat{T}_n = \frac{t_1}{1+C} [(n+0.5)^{1+C} - 0.5^{1+C} + (1+C)] \tag{8.10}$$

The difference between these two formulae is negligible: you may choose the one you prefer.

Figure 8.8 shows how the total cost grows for two values of $\lambda_C$, assuming $t_1 = 1$.

## What About the Average Cost?

The average cost is easy to compute, once the total cost is known. The average cost of manufacturing $n$ products, starting at rank 1, is given by:

$$\bar{t}_n = \frac{\hat{T}_n}{n}$$

## What About Production Lot?

Suppose you manufacture a lot of products, starting at rank $m$ and finishing at rank $n$. A good approximation, very easy to use, is given, as a result of (8.10), by:

$$\hat{T}_{m,n} = \frac{t_1}{1+C} [(n+0.5)^{1+C} - (m-0.5)^{1+C}]$$

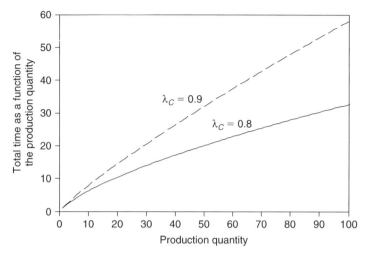

**Figure 8.8** Crawford's law: total time for $\lambda_C = 0.8$ and 0.9 with $t_1 = 1$.

### 8.2.3 Comparing Wright's and Crawford's Laws

Figure 8.9 gives the unit time curves for both Wright's and Crawford's laws, for the same value of the parameter: 0.9, starting from the same FPC of 1.

The curves look rather similar, but are not identical: for the same value of the learning rate, Wright's law always gives lower values (unit and total) than Crawford's law.

If you use always the same law, you will have no problem. But what about exchanging learning rates with people who do not use the same law as you do? You have to adjust the learning rate you receive to adapt it to the law you use.

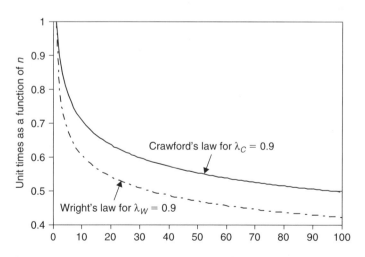

**Figure 8.9** Unit time given by Wright's and Crawford's laws for the same $\lambda = 0.9$.

Unfortunately, you cannot adjust it for both the unit cost and the total cost. Figure 8.10 illustrates:

- You are using Crawford's law with a learning rate $\lambda_C = 0.9$ (full curve); you get a total cost $T_{100}$ for a production quantity of 100 items, starting from a FPC $t_1$.
- Someone, who is using Wright's law, asks you what should be the value of the learning rate $\lambda_W$ he/she should use to get the same total cost $T_{100}$ from the same $t_1$.
- By trial and error using the previous formulae, you get $\lambda_W = 0.92163$. As Figure 8.10 shows it, you get the same total cost, but the unit costs are different; to get the same total cost, Wright's law starts with lower unit costs than Crawford's law, and, after rank 40, has to use higher unit costs.

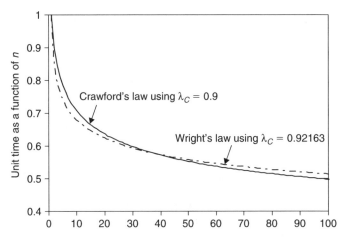

**Figure 8.10** Unit time given by Wright's and Crawford's laws for different values of $\lambda$ in order to get the same total cost for $n = 100$.

This example shows that, if both curves look similar on Figure 8.9, they are not identical. Furthermore and due to that, the change in the learning rates would be – slightly – different if the total quantity considered amounts to 100 or to 1000.

In conclusion:

Wright's and Crawford's laws give equal **total costs**, but **not for the same value of the learning rate!**

In order to get the same total costs, Wright's learning rate must be greater (closer to 1) than Crawford's learning rate.

**Unit costs can never be the same for a whole production.**

Consequence: if someone gives you the value of a learning rate, always ask which one it is! If you do not know, do not use the value: you can make big errors. For instance, with a learning rate of 0.9 and a FPC of 1, Crawford's law gives a total cost for 100 units of 58.14, whereas Wright's law gives for the same input 49.66. The difference is significant.

Be careful about comparing rates between European and American books if the law which is used is not given.

The choice of the law to be used is not important as long as you are consistent.

From now on, we will use only Crawford's law (but we will keep the symbol $\lambda_C$ in order to avoid any mistake).

## 8.2.4 In Practice

What value should the parameter have? The normal way to find the parameter value is by *trial and error on your own data*. Trial and error is necessary, because, as we mentioned it, you must eliminate the random fluctuations: although some mathematical technique could be used for doing that, we prefer the visual inspection on a graph.

Let us see on the example given in Figure 8.2. After some tests, we found that the set ($t_1 = 1.85$ and $\lambda_C = 0.85$) about correctly modelizes the first part of the data; Figure 8.11 illustrates.

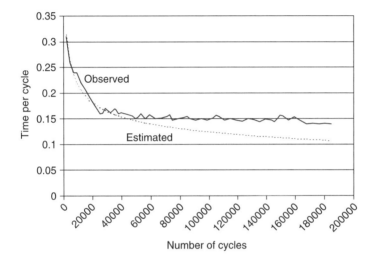

**Figure 8.11**  Observed time per cycle (full line) and estimated (modelized) time per cycle (dotted line).

This Figure 8.11 shows two points:

- The estimated line might be good enough (what we mean by "good enough" is explained on the general information about "what is the purpose of a model?") up to about 32 000 cycles. Random fluctuations are removed and the computed values might be, in this range, satisfactory for the decision-maker.
- But the model becomes less and less adequate as we go beyond 32 000 cycles.

This remark is about general and explains the success of the broken line in practice (see Section 8.3.1), as soon as the number of operations – or products – becomes large.

This is very good when you have your own data. Now, if you have no data, is it possible to propose a learning rate? This is very difficult and values you may find in articles should be used with caution, especially of course if the production quantity is large!

What can be given is some guidance. This guidance is based on the factors that may or should influence learning (as mentioned earlier, the values indicated here refers to Crawford's law):

1. The more complex a task, the less learning may be expected. This comes from the fact that the types of learning are not the same for simple tasks and for complex tasks, even if we use the same word to designate them. If the task is very easy,

learning means acquiring reflexes (remember the time it took for you to learn how to walk); when "learning" is finished, you do not think any more to what you are doing: it becomes "unconscious". This is certainly the case if you perform thousand times the same gesture which takes only a fraction of a minute; in such a case, learning rate between 0.80 and 0.85 might be observed (from our experience, it is extremely rare to observe values lower than 0.80 and we could add that the lower bound is far less likely to occur than the upper bound). If the task is more complex, learning does not mean any more acquiring reflexes, but a better coordination of gestures, avoiding small mistakes, etc.; in such a case, learning rates between 0.85 and 0.90 might be observed. Another way to say the same thing is: the shorter the task, the lower the value of the learning rate.

2. A study was made on gestures, based on the MTM (MTM is described in Ref. [7]). The result was that the learning was different according to the fundamental gestures involved in the activities. The gestures were grouped into three sets according to the speed at which the learning occurs: from very simple gestures (for which the learning is very fast) to more complex gestures. It logically found out that the duration of learning depends mainly on the size of the third group:
   - activities consisting of 70% of complex gestures needed 14 000 cycles to be about stabilized,
   - activities including of 40% of complex gestures needed 7000 cycles,
   - activities with only 10% of these gestures needed 900 cycles.

   This helps explain that different learning curves may be found in the same factory, depending on the observed activities.

3. C. Burnet (Ref. [16]), working on airplanes distinguishes two kinds of work: structure and equipment (he called "equipment" wires, pipes, furnishings and an array of small widely differing components) and found out that learning was faster on structures than on equipment. His interpretation was that structure was a much more repeatable work than equipment. It is interesting to note that he established – from historical data – that the learning was much faster on airplanes with a high ratio of structure weight to total empty weight (without the engines); this confirms his first finding. Surprisingly enough he observed $\lambda_C$ as low as 0.61 for an aircraft with very little equipment. The average value was about 0.77, which is already very low ...

4. The more automatized a task, the less learning may be expected: if the task is fully automatized, no learning at all is expected[11] ($\lambda_C = 1$).

5. As we will see in the next section, the higher the production rank, the lower the learning rate will be (with the exception of the very first units).

Two advices may be added:

- Be very careful about using learning rates you may find in production papers if you know nothing about the production organization, the nature of the tasks (activities or products?), the level of automation of the tasks, the relevant production rank and even maybe the type (Wright or Crawford) of the rate.
- Be prepared to observe learning rates in your own plant. From these observations, in relation with the duration of the tasks and the level of automation, you

---

[11] There is however, when the production starts a phase during which some learning is observed (though the name of "debugging" production procedures would be more adequate). This phase is dealt with under the name of "pre-production".

## 8.3 The Continuous Production: More Complex Models

The "pure" models we described do not exactly fit with the observations. The purpose of this section is to show how these laws can be a first step towards models which better fit with the reality; these models are generally more complex than the models already described.

The most frequent criticism of these "pure" models is that there is no limitation when the quantity becomes very large: the unit cost becomes smaller and smaller … This is in contradiction with the experience.

The reaction to this evidence is natural: let us complexify the models. Yes, but by how much?

This is a very general question in models building for which there is no definite answer. The general information page about "What is the purpose of a model?" brings some elements to the discussion.

### 8.3.1 The Broken Line Model

The broken line model is the first, and probably, the most understandable, complexification of the pure models. The logic is very simple: the pure models do not fit the actual data on a large range, but it seems that they may fit them by pieces.

Let us return to the example: the pure law with the set ($t_1 = 1.85$ and $\lambda_C = 0.85$) gives satisfactory and understandable results up to the rank 32 000 (about). From this rank, let us start with another set. In Figure 8.12, we use the following set from rank 32 000 and beyond ($t_1 = 0.35$ and $\lambda_C = 0.95$), also found out by trial and error.

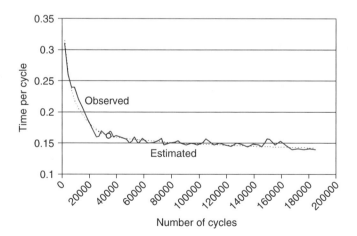

**Figure 8.12** Observed time per cycle (full line) and estimated (modelized) time per cycle (dotted line). The model changes at rank 32 000 (the circle).

## General Information: What is the Purpose of a Model?

A model is a representation of the reality. The main question about a model is: how accurate should the model be?

This question reminds us a story[12] of maps: in some kingdom, the king created a committee of geographers in order to draw a map of the kingdom. The first map was made in a couple of months; it was presented to the king who found it was not precise enough. The geographers started a new one: it took them a couple of years, before it could be presented to the king. This king found it was much better, but asked the question: Could you do it more precise? The geographers restarted their work; it took us a couple of decades. This time the map was perfectly accurate: every tree, every rock, every little path were represented on this map. The king was happy! But the map was really unusable: far too large, with too many details, it could not help anyone. The map finished abandoned to the rain and the wind.

This story is interesting because it helps explain that you cannot build a useful model if you do not decide first what is the purpose of it.

Any information in the domain of cost has just one purpose: helping a decision-maker. A model is just a tool for this objective. So the real question is: what does the decision-maker really need?

Let us try to answer this question.

The decision-maker needs a compromise between two objectives:

1. He/she needs of course a forecast of a reliable quality. Remember the discussion we had in Chapter 1 about "What is a good estimate?": the decision-maker needs an estimate good enough to base his/her decisions on. He/she does not need to get a figure with $n$ numbers after the decimal point: his/her focus is much more on the reliability of the estimate than on its precision. Even if the precision is modest (for instance 20%) he/she must trust it.
2. He/she must understand the logic of the model! The confidence he/she may have in a forecast is based on history (how often was the cost estimator right or wrong?) but also – and maybe it is more important – on his/her understanding of the model. In order to accept the output of a tool, human beings need to understand the logic of it.

We, as models builders, have the defaults of our qualities: we try to do the best – and we are often able to do it –, forgetting how our forecasts will be received, interpreted and eventually accepted or rejected. One of the most important factor is the capacity of the decision maker to understand the logic between the inputs and the outputs.

We should prefer simple models, even if their results are not extremely accurate, to complex models which may be very accurate but will not be relied on because only few people understand them.

It does not mean that models should be too simple: nature is complex and we have to modelize nature. But let us find out a good compromise.

We will return on this point in the part of this book dealing with conceptual models (Part III).

---

[12] We borrowed this story from someone else, but did not remember from whom. We hope the author will forgive us to mention the story without referring to him.

This time the model (which is now composed of two "sub-models") gives a satisfactory – and always understandable – response to the problem.

This model (the use of just two sub-models) might be sufficient in many cases. Sometimes a more complex one is necessary. Table 8.1 illustrates what was observed[13] in the aircraft industry.

Table 8.1 Learning rates (Crawford) observed in the aircraft industry.

|  | $\lambda_C$ | From rank to rank |
|---|---|---|
| Mechanical parts | 0.92 | 1–6 |
|  | 0.84 | 6–60 |
|  | 0.95 | 60–130 |
|  | 0.97 | >130 |
| Sheet-iron works and sub-assemblies | 0.88 | 1–6 |
|  | 0.82 | 6–60 |
|  | 0.90 | 60–150 |
|  | 0.95 | 150–200 |
|  | 1.00 | >200 |
| Assembly – fitting out | 0.85 | 1–6 |
|  | 0.76 | 6–60 |
|  | 0.85 | 60–150 |
|  | 0.90 | 150–200 |
|  | 0.95 | 200–400 |
| Integration | 0.85 | 1–6 |
|  | 0.76 | 6–60 |
|  | 0.85 | 60–150 |
|  | 0.90 | 150–200 |
|  | 0.95 | 200–300 |
|  | 1.00 | >300 |
| Control | 0.94 | 1–6 |
|  | 0.86 | 6–60 |
|  | 0.95 | 60–150 |
|  | 1.00 | >150 |
| Global | 0.88 | 1–6 |
|  | 0.80 | 6–60 |
|  | 0.92 | 60–200 |
|  | 0.99 | 200–400 |

Let us discuss these values. Three phases can clearly be distinguished in the learning process.

1. The decrease in the production cost is low for the first six ranks: we can call this phase[14] the "*discovery phase*". It is the period of time during which the worker discovers what he/she has to do; he/she has to read the documentation, understand it, reread it, adjust the work place, sometimes debug the procedures, etc. until he/she becomes familiar with the operation. During this period of time, first the decrease is low[15] due to the time spent in auxiliary activities, second the level of

---

[13] Source: Jean Chevalier; He was employed as an engineer in economics by CNES (the French space agency). Jean works now as an independent consultant for the industry.

[14] We will meet again this phase in Part III about conceptual models: it will then be called the "pre-production phase".

[15] Some authors even say there is a "negative learning" during this phase; they explain that by a lot of debugging and modifications. This may happen if production was not prepared with enough details.

fluctuations between two consecutive ranks is rather large (do not expect to see a clean change in the manufacturing cost): the learning rate is therefore an average.

2. From rank 6 or 7 up to rank 60, we observe the "real" learning curve (the values suggest that the activities were fully manual (no automation at all)) as it was described by Wright and Crawford. Let us call it the *"true learning phase"*. The duration is much higher: it explains why it was discovered first. We can also add that the phenomenon is much clearer than during the preceding phase: it is quite common to look, in the industry, at very nice and proper learning curves, with very little fluctuations.

3. The third phase is called the *"stabilization phase"*: after rank 60 (about) the learning rate becomes closer and closer to 1, which means the learning does not happen any more because the process is more or less stablized: there is not so much possibilities of learning. Depending on the operation, it is quite possible to still observe a small decrease of the time spent (or the cost): the stabilization does not happen suddenly and this explains why you can see, in the table, this phase us split into two or even three sub-phases, finishing with a learning rate equal – or nearly equal to 1. It would of course be possible to write down a new formula taking this slow stabilization into account, but it would be pure cosmetics: it is easier to adjust the learning rate.

Nobody can tell exactly where the change of phase occurs: Is it 60 or 61 for instance? But this is not important and will not make such a difference (one can always draw manually a soft transition between the phases). The important thing is the concept: being warned, the cost analyst will easily recognize the phases (the human brain perceives only – the exceptions are the true researchers – what it is trained to look at) and will be able to use them.

Using such a table requires a little of attention because all the three phases should be adjusted (each phase will have its own $t_1$). As an example we reproduce on Figure 8.13 the learning curve for mechanical parts. We used, for drawing this curve, the following values of $t_1$ depending on the phases: 1, 1.42, 0.685 and 0.593 (found by trials and errors) in order to make the curves joint gently.

The reader can easily see the changes in slopes at least for ranks 6 and 60 (the change at rank 130 is more difficult to perceive). The apparently sharp change during the first phase is due to the fact that we use the beginning of the learning curve: the reader has already observed that the decrease is much higher at the beginning of the curve than later on.

For the reader who is familiar with the log–log scales, we reproduce in Figure 8.14 the same curve with these new scales.

### 8.3.2 Other Laws

Some people proposed different laws to solve the problem of the cost trend towards 0 we saw with the "pure" laws of Wright and Crawford. Let us just mention two of them.

**De Jong**, for instance, proposed (in 1952) a law which can be considered as a substitute of the "broken line model"; this law was based on observations on short activities (less than 1 min). As can be expected with more complex laws, this law needs two parameters, plus of course the variable $t_1$; he wrote:

$$t_n = t_1[a + (1-a)n^C]$$

# The Cost Improvement Curves

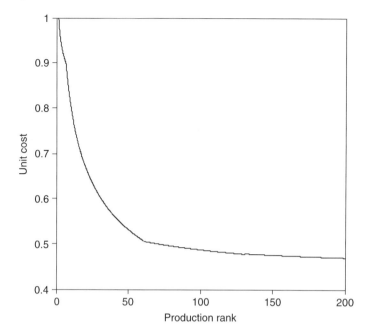

**Figure 8.13** Example of the broken line model (linear scales).

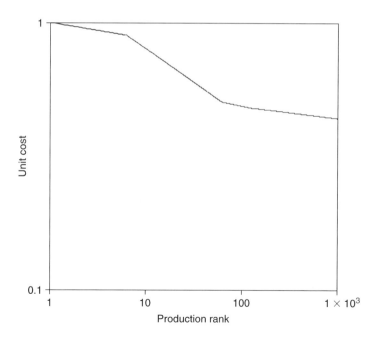

**Figure 8.14** Example of the broken line model (log–log scales).

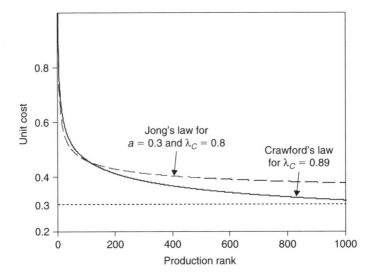

**Figure 8.15** Jong's law compared to Crawford's law.

It is clear that if $a = 0$, then this law becomes Crawford's law and that when $n$ (the number of activities or products) becomes very large, then $t_n \to a : a$ is then the minimum value that the time and the cost can have, as illustrated on Figure 8.15 (drawn with $t_1 = 1, a = 0.3$ and $\lambda_C = 0.8$).

For obvious reasons, Jong's law always gives unit cost lower than Crawford's law. Consequently, if we want to get the same cost for the first activities or products, we need to use a Crawford's law with a $\lambda_C$ higher than the one used for Jong's law, as it is the case in Figure 8.15.

This curve is very rarely used; but it could be useful for very large quantities.

**Lazard** proposed (1963) to use an hyperbola with an horizontal asymptote, corresponding to very large production ranks.

## 8.4 More Complex Situations to Modelize

Wright's and Crawford's laws were built for forecasting the cost of continuous productions of a new product, other things (technical definition, tooling, man power, production rate, …) remaining unchanged. In real life, things may change. This section illustrates how these laws can still be used to modelize more complex situations (once again, the illustrations use Crawford's law on unit times).

### 8.4.1 Modifying the Production Methods

Let us assume that the purpose of the modification is to decrease the manufacturing time. We can also assume that the learning rate remains unchanged (this is generally true if the modification is limited).

The modification occurs at a given rank: let us call it $m$; at that rank, the manufacturing time would have been $t_m$ if no change has occurred. It is realistic to consider that:

1. only a portion $p$ of this time is concerned by the change;
2. that, on the new portion $p$, learning restarts as if it were rank 1 instead of $m$.

If we assume that the modification reduces the manufacturing time of the new portion by a factor $\alpha$ ($\alpha < 1$), then the total manufacturing time at rank $n$ ($n > m$) will be the sum of two terms:

$$t_n = t_1[(1-p)n^C + \alpha \times p(n-m+1)^C]$$

To illustrate what happens, let choose the following example (Figure 8.16):
- $t_1 = 1$ (the FPC),
- $C = 0.85$ (the learning rate, supposed to be remain unchanged),
- $m = 20$ (rank at which the modification occurs),
- $p = 30\%$ (portion of the time concerned by the change),
- $\alpha = 0.7$ (reduction factor introduced).

Figure 8.16 shows that, at rank 20, there is a small "loss of learning" until the time reduction can clearly happens.

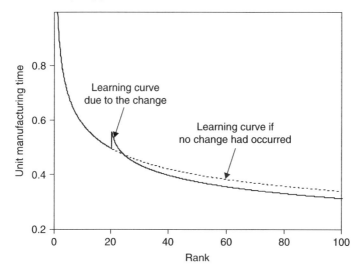

**Figure 8.16** Modification of the production methods.

The same logic can be easily applied to situations such as the modifications of the product by adding something, removing something or transforming something inside the product.

### 8.4.2 Interrupted Production

When the production is interrupted for some time, it is difficult to expect that, when it restarts, the manufacturing cost – or time – will restart exactly at the value it had

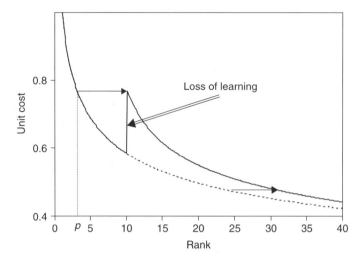

**Figure 8.17** Interrupted production.

when the production stopped: some learning is certainly lost due to the change of personnel, new supervisors, the production of new organization, etc.

The basic idea of the authors who wrote about the subject is illustrated on Figure 8.17: starting with the curve observed before production stops (the first lot), they modelize the curve when production restarts (the second lot) by using the same curve, but shifted on the right, as if the production would restart at rank $p$. In the figure, production stops at rank 10; when it restarts, instead of restarting at the cost of rank 10 (about 0.6), it really restarts at the cost measured at rank 3 (about 0.77). The difference between 0.6 and 0.77 is due to the loss of learning; it is measured by the ratio of the actual loss to the maximum loss (restarting at $t_1$) which would be here 0.4: the loss of learning is 0.17/0.4 = 0.425 or 42.5%.

The problem is: how can we forecast this loss of learning? Obviously, we have to take into account the duration during which production stops.

A. R. Ballman (Ref. [6]), from a paper published by Anderlohr (Ref. [2]), proceeded in two steps:[16]

1. First of all he distinguished three major factors which should be taken into account: in-plant factors (supervision, personnel, tooling, production continuity, methods, configuration), availability of personnel, retention of specific knowledge; he then draw curves on how these factors change with time. Giving "weights" to each factors, plus the way they change with time, he eventually modelized the loss of learning.
2. He compared the result with data and adjusted the formula. Here is about the formula he proposed (the "duration" is the number of months elapsed between the stop and the restart of production):

$$\text{loss} = 1 - 0.8 \times \text{duration}_{\text{months}}^{-0.3}$$

---

[16] This way of doing: "theoretically" establishing a model and then adjusting it from a few observed data might be a way – unconventional – to build models when the number of data is limited. The "theory" gives the major trend and the data allow to correct it.

For instance, after 6 months, the modelized loss of learning 0.53 (or 53%); after 12 months, 0.62. But he admitted that the dispersions, limited for the first 6 months, tended to increase as time progressed. His conclusion was then, that, for practical purposes, "costs tend to recede to near the value of unit one on the learning curve after about 1 year".

Generally speaking, when we have to normalize data or denormalize forecasts, a linear loss of learning from 0% to 100% generally gives good results when production stops from 0 to 12 months. It is not very precise – and cannot be when you consider all the factors which influence the loss of learning – but sufficient in practice.

## 8.5 Applications

The purpose of this chapter is to normalize data and to denormalize forecasts. If you use a conceptual model, it is very likely that these processes are built in (they are in most of these models) and the purpose of this chapter is only to explain how it does.

If now, you want to create your own statistical models, you will have to use yourself the logic developed in this chapter.

### 8.5.1 Normalization

Suppose you receive a cost value corresponding, for a product or an activity, to a given rank – or to a given lot starting at rank $m$ and finishing at rank $n$.

You want to compare this cost with other costs you get for other products, for which you do not know the cost at the same production rank. In order to be able to compare these costs, you must normalize them. If you do not do that, you saw in this chapter that the costs at various ranks can be completely different and you will probably not be able to create your own model.

For doing this normalization, you have three decisions to make:

1. Select the **model** you will use. In this chapter several models have been presented and discussed. They do not cover all the researches which have been done on the subject (it was not its purpose), but they illustrate the relative complexity of the problem: selecting a model implies to get some understanding about the industrial organization.
2. Enter the value of the **parameter**(s), which is for instance the learning rate. This really implies that you know something about the product (is it complex or simple?) and about the level of automation that will be used, unless you were able to measure it directly in your own plant.
3. Decide at what **rank** you want to normalize. Most often the rank selected by the cost analyst depends on the production size:
   – For small production, people choose rank 1. This quite feasible, but it should be recalled that the costs of the first ranks may present an important dispersion: the cost at rank 1 is really a "theoretical" FPC.
   – For medium-size production, one generally prefers a more stabilized rank. For instance the aircraft industry generally chooses rank 100.
   – For mass production, a much higher rank can be chosen (1000 is not rare): the principle is to wait long enough until the debugging of the production procedures is nearly finished.

### 8.5.2 Denormalization

The model – unless the model you use has built-in algorithms to take into account the learning process – will compute for you a cost at the rank you normalized the data you used to built it.

Now you may want a cost at another rank, or a total cost for a lot, or a cost taking into account a production break, etc. You have to denormalize this forecast.

The logic is, for the first two steps, the same as the one explained in the previous section: model selection, parameter(s) choice. Preferably, you will use the same ones as those you use for normalization. Once this is done, compute with the ranks you are looking for, using the formulae developed in this chapter.

## 8.6 Conclusion

Learning is not an academic feature, it really exists in the industry.

Many things have been written on the subject of "learning" and its causes. But the influence of each individual cause is still very difficult to assess.

Experience remains necessary ...

All the models we presented in this chapter are "parametric" models. The process of building a model – here: looking at the data, selecting an idea (what the model could be), adjusting it with parameters – should start to be familiar to you.

# 9 Plant Capacity and Load

*By Gilles Turré in charge of estimation methodologies at PSA (Peugeot-Citroën) Cars Manufacturer, France*

## Summary

This chapter deals with "mass production". Mass production means that the same product – for instance a car – is manufactured in very large quantities per year, the million being an order of magnitude.

The subject of plant capacity is therefore a very important one for the cost analyst, as – and this is a very familiar subject – the cost of a product very strongly depends on the produced quantity. This person should then being able to answer the first question: How the cost of the products is going to change when the plant capacity does change?

Unfortunately no plant uses its full capacity all the time: the load (the amount of the capacity which is actually used) depends on the demand for the product and the plant manager must all the time adjust the production to the demand. It is not difficult (the mere existence of the fix costs – such as the depreciation of the investment – is only one side of the subject) to understand that, for a given capacity, the cost of the products will strongly depend on the load. Therefore the second question which can be asked to the cost analyst is: How does the cost of the products is going to change when the plant load does change?

Once this is done, both subjects have to be simultaneously taken into account.

This chapter deals with tangible products, having a physical existence. However, we have the firm conviction that all the rules established between size, capacity, load and the cost are applicable to the services and any other immaterial productions.

Simply, theory had been established earlier, with numerous examples, in the domain of physical objects. We are sure that the reader will be able to make easily the pertinent analogies.

## 9.1 A Few Definitions

### 9.1.1 What Is a Product? What Is its Size?

*A product is a thing*, as a tank, a building, a watch, a car, a plane, a hole in the soil. In these cases, the size is usually expressed in volume, surface or mass terms. A tank of 5000 gallons, a building of 50 000 m$^2$, a hole of 500 000 m$^3$, a car, a plane or a watch of 1000, 10 000, and 0.01 kg.

In the following, we shall deal with:

- The total cost of the product. This total cost is the cost of the whole product, the car or the building, for instance. This cost will be called the unitary cost and will be represented by $Cst_u$.
- The specific cost of the same product. The specific cost is the cost of one unit of size, a kilogram of car or a square meter of building, for instance.

*Some products are less tangible*: sporting, cultural, commercial events, services of all kinds, may be considered as products of a project, and therefore as objects of an estimate for which the size, in terms of spectators, or duration, or surface, makes sense.

*A more complex case is the production of products*. The project of making a new car leads to a new product, which has its own characteristics of size (length, surface, mass, etc.). But it leads also to a new *production* which is the construction of a given number ($Qty_{actual/yr}$) of this product in a given place, during a given duration (e.g. Corvette, 4 800 000, 8 years, Detroit, MI, USA). The cost of this production is symbolized by $Cst_{t/yr}$. The main characteristic of size of this production is the number of products (for cars and planes, for instance, expressed usually in units) or the number of products times the size of the products (for boat construction, for instance, expressed usually in tons).

Therefore, it is important to notice that the *total cost of the product "car"* ($Cst_u$) is in fact the *specific cost of the product "production"* ($Cst_{t/yr}$).

It is a fact of usual observation that the specific cost of one object (one gallon of a tank, for instance) decreases when the size of the object (the tank) increases. It is also a fact of usual observation that the cost of an object of mass production (a car, for instance) decreases when the size of the production of this object (the whole stock of this model) increases.

The behavior of the cost with the size seem really follow the same laws for single products and for mass production. It is only important to remember that a product of mass production must never be considered alone, as a single product, but as an unit of another product, which is its whole production.

### 9.1.2 Parameters of a Production

This chapter uses several notations which are not necessary in the other parts of the book. They are described below:

*Total Production, Items and Equivalent Items: Period and Size*

The production of a plant consists of large quantities of "items" (or parts, or equipments, or sub-systems or systems, the word "item" being the generic term). This chapter considers:

- Either the total production during a **period** (generally the year). A plant may produce different types of items (small or large); in order to be able to add these items together, we define a "reference" or "equivalent" item and each item is considered as a given number of this equivalent item. The quantity produced per year $Qty_{yr}$ adds "equivalent items" together.
- Or a particular item. The term "unit" (for the cost for instance) refers to this particular item.

"*Size*" is a generic term, when referring to the total production the size is (also sometimes called the production "volume") equal to $Qty_{yr}$. When referring to a particular item, the size is often quantified by its mass.

## Plant, Capacity, Load and Production Rate

A plant is designed for producing during the year a given quantity of "equivalent items" $Qty_{yr}$. This given quantity is called the "capacity" of the plant. Several capacities may be defined, such as theoretical maximum $Qty_{th.\_max/yr}$, or practical maximum $Qty_{pr.\_max/yr}$. This capacity is called "non-particular"; when referring to a particular item, the production capacity, defined by the tooling dedicated to this item is called "particular".

The load of the plant quantifies the usage of the capacity which can be expressed as:

$$\text{Load} = \frac{Qty_{yr}}{Qty_{pr.\_max/r}}$$

The production rate is the quantity produced per unit of time, it is naturally called $Qty$/time.

## 9.1.3 Plant Capacity: Dedicated and General Capacities

The term capacity immediately makes think of the volume of a tank. It is an analogy which is relevant when one speaks about the capacity of a foundry furnace or a distillation tower, which are, indeed, large cans. It is less relevant when one speaks about the capacity of a stamping line, molding line or assembly line.

For a plant, the capacity is always quantified by a producible quantity per year (if – as we do it in this chapter – the period is chosen as the year). There are several measures of this quantity.

### Capacity Defined by the Volume
Actually, in the case of the foundry furnaces, the heat-treatment ovens, the mixers and generally speaking in all the production techniques of which potential is a cubic measure, we note no fundamental difference in the design of the machines, even if we multiply the capacity by an important coefficient, such as 10 or more.

### Capacity Defined by the Production Rate
Let us take several examples, starting with the stamping lines.

In the case of a stamping line, we note important differences in the design of the line and its machines with the production rate: small rates (less than 200 pieces per day) are done on presses with manual handling, average rates (more than 200 pieces per day) on presses with automatic handling between presses and manual loading and unloading of the line, high rates (more than 1000 pieces per day) on lines with automatic loading and unloading, and very strong rates (more than 5000 pieces per day) on transfer presses. Let us add that it is very rare that the production rate of a product is sufficient enough to monopolize a line of presses. Such a line is therefore usually designed for manufacturing several products; this implies that one has to change the set of the specific tools, set which usually exists in one specimen only.

About the Injection Molding Lines   In the case of the molding by injection of plastic parts or aluminum parts, the need of cooling time makes these technologies intrinsically slower than stamping. It is not rare to see one or two machines monopolized by the production of the same part all year long. These machines remain nevertheless general-purpose machines which will be able to pass to another production after extinction of the product that they manufacture. The tool they support (mould) is, on the opposite, completely specific. It must be manufactured in as many specimens as the ratio "maximum rate of the mould/manufacturing rate" indicates.

About Machining Lines   In this case, we notice now other differences. Whereas the stamping or molding workshops are divided into homogeneous sections (large presses, small presses, cutting), the current machining workshops (it was not the case 30 years ago) are generally organized in blocks dedicated to the manufacture of one product or one family of products. They include machines in the quantity and quality required to ensure manufacturing on line without intermediate storage (or strictly limited to the management of the micro-failures).

About Assembly Lines   Last but not least, assembly lines always are, and always were, organized by product families. Exclusively manual at the beginning, they tend to be mechanized and look more and more like machining lines. Nevertheless their design is always based on a sequence of manual operations needing a certain time, uncoupled (off line operations) or in series (on line operations). A line treats a product or a family of products. In this last case an average time is allocated to the operations, and it matters that the products follow one another so that a compensation is permanently established between complex products and simple products; otherwise the allocated time will be temporarily insufficient and the line is disorganized.

Basically, assembly lines are versatile and re-engineerable, which means that they accept different products in the same production flow, and that they can be easily adapted to a modification of the characteristics of the products to be manufactured.

## *Dedicated and General Capacities*

The concept of capacity is applicable at various levels: we can speak of the capacity of a factory, of a workshop, of a line or of a tool set.

As the same plant can simultaneously produce several different items in different quantities, it is important to differentiate the capacity proper to a given product, the particular capacity and the non-particular capacity which is not dependent on a given product:

- The dedicated capacity is determined by the tools (more exactly the couple "material + tools", in the case – rare – where several materials can be machined by the same tools). The tools were designed for a given rate corresponding to the need for this product. We can assimilate a specific organizational block to a tool.
- The general capacity is the one which is determined by the general means of the workshop or the factory. It is always higher than the particular capacities individually taken, but can be lower than their sum.

## 9.1.4 Practical Summary of the Capacity Parameters

### *Instantaneous Rate: Bottleneck*

A work plan always involves a sequence of operations of which each one has a maximum instantaneous rate, and each one of these rates is likely to be expressed in different units: Fusion 30 ton/h, plate model 10 s per mould, moulding 2.5 min of cooling per moulded kilogram, stamping 40 strikes of press per minute, assembly 40 cars per hour, etc.

It is the slowest operation, known as bottleneck operation, which determines the maximum instantaneous production rate of the line $Qty_{max}$/time.

### *Opening Time*

A production line can be operated during a given time, expressed in hours per day and days per year. The maximum operation time is determined by the needs for programmed maintenance $Tm$. It is the operation requiring the longest programmed maintenance time which determines the maximum theoretical working time $Tw_{max}$, the unit being for instance the hour.

The working time is always defined as a fraction of a period. In the rest of this chapter this period will be taken as the year – but it could be the day or the month.

The quantity theoretically produced during this period is called $Qty_{th.\_max/yr}$.

### *Incidents, Failures, Breakdowns*

A production line is subject to stops and breakdowns. The incident rate $Ri$ is the ratio of the duration of the time of the stops for incidents to the total time of operation. The maximum practical time of work $Tw_{pr.}$ (also given in hours) is the theoretical maximum working time multiplied by the availability rate $1 - Ri$:

$$Tw_{pr.} = Tw_{max} \times (1 - Ri)$$

### *Scrap*

A part $Qty_{scrap}$ of the total production quantity $Qty$ must be rejected as scrap. The remaining production is the conform production. The scrap rate $Rs$ is the ratio $Qty_{scrap}/Qty$.

The conform rate is $Rc = 1 - Rs$.

### *Theoretical and Practical Capacities*

The theoretical maximum capacity $Qty_{max/yr}$ is the product of the maximum instantaneous rate $Qty$/time by the theoretical maximum working time $Tw_{max}$. It is a limit which is never reached.

The practical capacity $Qty_{pr./yr}$ is the product of the maximum instantaneous rate $Qty/\text{time}$ by the practical maximum working time and the conform rate:

$$Qty_{pr.\_max/yr} = Qty/\text{time} \times Tw_{practical} \times (1 - Rs)$$
$$Qty_{pr.\_max/yr} = Qty/\text{time} \times Tw_{max} \times (1 - Ri) \times (1 - Rs)$$
$$Qty_{pr.\_max/yr} = Qty_{th.\_max/yr} \times (1 - Ri) \times (1 - Rs)$$

## Output

We shall call maximum practical yield of an installation the ratio of the practical capacity to the theoretical capacity:

$$\frac{Qty_{pr.\_max/yr}}{Qty_{th.\_max/yr}} = (1 - Ri) \times (1 - Rs)$$

It is a maximum, because the yield does vary with the load of the installation (see below).

## Retouching

A part $Qty_r$ of the production $Qty$ can be neither good, nor scrapped, but sent to retouching. This part of the production will generate costs of final retouching, which will increase the production cost without changing the output of the line (in volume terms: the yield is not affected; of course this is not true in financial terms). Let us name retouch rate as the ratio of $Qty_r/Qty$.

The retouched production $Qty_r$ does not come in deduction from the conform production. It is a part of it. An excessive growth of the retouch rate can decrease the rate of the line of production, because the retouch line is saturated. But it is normally a transient situation.

For cost control purposes, one often defines a "direct conform production" $Qty - Qty_r$ to which a rate can be attached.

## Types of Installations: Homogeneous Workshops, Lines, Blocks

**In Homogeneous Workshop (Milling, Plastic Injection)**  It is the traditional organization: the workshops gather machines of the same technology. Each machine is independent of the other (they are not on the same flow of manufacture). Each one is surrounded by secondary equipments. These secondary equipments have a superabundant capacity in order not to limit the production of the main machine. The personnel who operate the machines are skilled and specialized in the shop technology (turners, milling-machine operators, grinding-machine operators, etc.).

**In Line (Assembly Line, Car Parts Machining Lines)**  It is the mass production organization: stations using different technologies, of about equivalent duration, are crossed by the same flow of products. Some operations can be optional. The operators have no particular qualification. Only the people in charge of the maintenance are skilled. Buffer stocks make it possible to uncouple the operations in order to avoid that a micro-breakdown on a machine causes a general stop of the line. The line finishes where the flow divides to feed other units (if it feeds one only, we can consider that the line carries over, even in a different location). The line finishes also when the flow arrives in an important enough storeroom to disunite the line from the installation which follows it.

In Blocks (Assembly of Sub-Systems) It is the recent trend: the blocks are an intermediate organization between the line and the shop. A small group of machines, or a principal machine and its additional equipments, are specialized in a given group of operations. The machines which make it up may not belong to the same technology; the block is not as "homogeneous" as the workshop. But, at the opposite of it, the machines are dependent and operate in sequence. A block can feed several lines, or other blocks. It can be itself fed by several others blocks. A block is a kind of short, specialize line. The blocks are not set out in a single flow, like the line, but in a network. Skills of the workforce are standard, similar to the line.

### Consequence on the Expression of the Capacity

A machine, a line, a block, whatever their size, have the same expression of capacity. It is neither possible, nor useful, to subdivide them.

The fact that it is not possible to stock an appreciable extent inside a line makes that it is necessary to look at it as only one machine. Buffer stocks which exist in the line do not make it possible to divide it, but they make it possible to avoid that the micro-stops cause the breakdown of the whole line. Therefore, these stops are not to be taken into account as incidents.

### Dedicated and General Capacities

An installation which receives tools always has a proper maximum rate higher than that of the tools it receives. A stamping press can, for example, strike up to 60 blows per minute, but the tool will not accept more than a given rate (25 blows/minute for the part A and 35 for the part B). A press for plastic injection, as another example, has a proper rate given by its capacity of plasticization and its speeds of opening and closing, but the rate of the moulding is mainly given by the time of cooling of the polymer in the mould, itself even dependent on the nature of the polymer and thicknesses of the part.

The proper instantaneous maximum rate of the installation is called non-dedicated capacity, or general capacities.

The instantaneous maximum rates imposed by the tools and the material are called dedicated capacities.

### Dedicated and General Capacities: The Case of the Blocks

A specific block in a workshop (e.g. of machining) has a proper capacity $Qty_{pr.\_max/yr}$.

A machining workshop composed exclusively of specific blocks theoretically has for its capacity the sum of the capacity of its blocks. Very often this is not the case, because the utilities, maintenance and the personnel do not make it possible to reach this total. The maximum which can be reached is the non-specific capacity of the workshop. For example a workshop composed of 10 specific blocks able to produce each 1000 parts per day can have a total non-specific capacity of 7000 parts per day: opposite to the case of the plastic injection press, the non-specific capacity is lower than the sum of the specific capacity.

## Load

The load of an installation or a plant[1] is the ratio of its production $Qty_{actual}$ to its capacity $Qty_{max/yr}$. According to the capacity outputs to which one refers, one may define several loads:

- load on theoretical capacity,
- load on practical capacity,
- load on nominal capacity.

The load on nominal capacity uses a conventional reference capacity which can be much lower than the maximum capacity. It is what is used in worldwide comparisons made by "Harbor"[2] between the manufacturing units of cars (the Harbor reference is the final plant working 250 days/year in two teams of 8 h). The load can thus easily exceed 100%.

The load on practical capacity exceptionally exceeds 100%.

The load on theoretical capacity, by definition, is always lower than 100%.

### 9.1.5 About the Cost

*Some Definitions*

We call "cost" here the full cost of a product at exit of the factory working as a home-worker. This means that cost includes beyond the direct cost:

- The depreciation of the machinery (we do not limit this presentation to the variable costs). The cost of the plant is an element of the production cost which is introduced, via the depreciation, in the labor rates for the buildings and for all the general-purpose equipments. Specific tooling – dedicated to a particular item – is not included in the labor rates: it is a direct cost of the items produced.
- The undivided expenditures of an autonomous factory (administration, restaurant, local taxes, etc.). We do not limit ourselves to the expenses of the workshop.
- The plant management and, more generally, the cost of everything which is strictly necessary to produce.

---

[1] The maximum capacity to take into account for the calculation of the load of a factory composed of installations receiving tools (moulds, stamping tools) is a delicate concept to define. Indeed, if one bases oneself on the non-specific capacities, the maximum capacity will be very high, and will be considered unrealistic by the experts. If one bases oneself on specific capacities, then it is necessary to define a "mean reference tool" which is a very conventional concept, therefore prone to interpretation and dispute.

[2] Harbor: company publishing a report of benchmark between the main car building plants in the world. It is brought to define normal conditions of use.

This also means that the cost does not integrate:
- The expenditure of research and development. We limit ourselves to the production costs.
- Commercial costs.
- *General and administrative expenses (G&A)*: headquarters expenses, especially financial expenses, research and development.

This full cost will be designated by the variable name *Cst*, an index being used for differentiating all the costs which are the subject of this chapter. (NB: R&D expenses, commercial costs may, in another context, be regarded as production expenses of commercial or technical solutions. The techniques described under may be applied to them, up to a certain extent.)

## Cost Parameters

Cost refers:
- either to the total cost of the whole production during a year: $Cst_{t/yr}$,
- or to the cost of one item: it is called unitary cost or $Cst_u$.

The "specific" cost is the cost per unit of size: it is then the cost of one equivalent item when we speak about the whole production:

$$Cst_u = \frac{Cst_{t/yr}}{Qty_{yr}}$$

or the cost per kg $Cst_{sp}$ when we speak of one particular item $Cst_u$/mass.

When referring to a particular item, several costs can be defined:
- cost of the material $Cst_{u/mat}$,
- cost of the work force (or "touch labor") $Cst_{u/wf}$,
- the part of cost which is sensitive to the plant capacity $Cst_{u/capacity\_effect}$,
- the part of cost which is sensitive to the plant load $Cst_{u/load\_effect}$.

Both $Cst_{u/capacity\_effect}$ and $Cst_{u/load\_effect}$ are important values. These are the values used in the process of normalization.

## 9.2 Cost of Production as a Function of the Size

This section is just an introduction to the general subject of the change of the products costs according to their size, subject which will be used in the rest of this chapter.

### 9.2.1 The Power Law or the Chilton's Rule

The power law (called the "multiplicative" law in Chapter 6 of Volume 2) gives a general form to the equation of the cost of an object according to its size, as well as the cost of a full production according to the plant capacity.

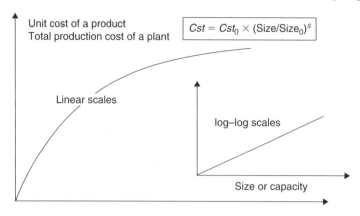

**Figure 9.1** Cost of function of a product size or of a plant capacity.

It comes from very usual observations that the cost of an object is less than proportional to the size of this object. It is true, for instance, for an engine according to its power, or for a building according to its floor surface, or for a ship according to its payload.

This is proved to be correct, all other things being of course equal. For example for flats located in equivalent districts, for ships transporting the same type of products, for engines designed for the same use and produced in similar quantities.

This is proved to be correct within reasonable limits: if the size of a mechanical unit becomes very tiny, one enters the field of the watch making industry. In this field, the effect of size is negligible in front of the increasing complexity the size reduction causes. If this size becomes colossal, one enters the field of the exceptional one, which by definition, does not know laws.

The form of the power law is:

$$Cst = a \times Size^s \quad \text{with } s < 1$$

The size being the product size or the plant capacity. The graph of this relationship is given on Figure 9.1 in linear and log scales.

This "power law" is used daily in engineering calculations, especially in oil and petro-chemical engineering.

It is only an economic model, and therefore not a physical law provable through experiments. Empirically one can at least observe that it accounts for the noted phenomena far much better than the competitor models, such as the linear (cost = $a \times$ size + $b$) or exponential models.

The practical significance of the power law is that when the size increases by $x\%$, then the cost increases by $y\%$. One can demonstrate that the exponent $s$ is given by:

$$s = \log(1 + y\%)/\log(1 + x\%)$$

The power law is linearizable to the form:

$$\log Cst = s \times \log Size + \log a$$

### Chilton's Rule

The value of the exponent $s$ is given by experience. If it does not exist, and if the subject of calculation belongs to the usual engineering field, Chilton proposed to take $s \approx 0.6$.

# Plant Capacity and Load

**Table 9.1** Extrapolation exponents $s$ used in petroleum engineering[A] (Ref. 10).

|  | $s$ |
|---|---|
| *Refinery units* | |
| Vacuum distillation | 0.71 |
| Distillation | 0.67 |
| Hydrogen units | 0.65 |
| Sulfur units | 0.60 |
| *Equipments* | |
| Furnaces | 0.85 |
| Compressors | 0.82 |
| Strippers | 0.73 |
| Balloons | 0.65 |
| Exchangers | 0.65 |
| Centrifuge pumps | 0.52 |
| Boilers | 0.50 |
| Storage tanks | 0.35 |

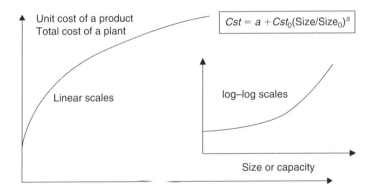

**Figure 9.2** Cost as a general function of product size or of plant capacity.

The above equation becomes (it is known as the "Chilton's law"):

$$Cst = a \times Size^{0.6}$$

Table 9.1 gives the exponent observed in the petro-chemical industry. It is interesting to notice that:

1. the exponent is nearly always less than 1,
2. the more complex an equipment, the closer is the exponent to 1: simple equipments have an exponent in the vicinity of 0.5, whereas complex equipments are characterized by an exponent in the vicinity of 0.7. The value of 0.6 mentioned earlier is therefore an order of magnitude.

## Advanced Power Law

This form makes it possible to account for fixed costs (cost which are independent of the size). It is therefore more general and allows modeling of most cases. Its form is given by (Figure 9.2):

$$Cst = a + b\, Size^s$$

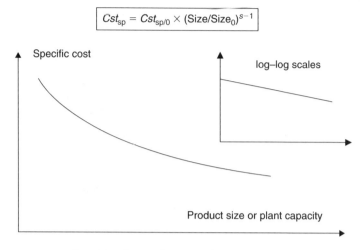

**Figure 9.3** The specific cost as a function of the size.

It is unfortunately not linearizable, which may (but should not) restrict[3] its use.

Experience shows that for large equipments, such as those listed in Table 9.1, the first part of the formula is very small compared to the second part: it can therefore be neglected, as we did for the Chilton's rule.

### 9.2.2 The Power Law for Specific Costs

We have considered above *the full cost Cst*, that is the cost of the whole "object" (which can be a car or the production of thousands of cars). If now we consider the cost per unit of mass (for a car), or volume (for a "volume of cars") or any other measure of size, we can define a specific cost $Cst_{sp}$. Note that the specific cost of the full production is equal to the unit cost $Cst_u$ of an item.

$Cst_{sp}$ varies as the following formula, where $Cst_{sp/0}$ is the reference cost for the size $Size_0$:

$$\frac{Cst_{sp}}{Cst_{sp/0}} = \left(\frac{Size}{Size_0}\right)^{s-1}$$

For instance, if the cost of a petroleum storage tank will vary according to Figure 9.1, with an exponent $s$ of 0.35, according to Table 9.1, then the cost *per gallon* of the storage will vary according to Figure 9.3, with an extrapolation exponent $s - 1 = -0.65$.

## 9.3 Relationship Between Plant Size and Costs

### 9.3.1 Cost of a Production as a Function of the Plant Capacity

This section deals with the production cost of a plant.

---

[3] the way such a relationship is created is explained in detail in Chapter 3.8.

## Size of the Plants

The production cost depends on the plant size, all other things being equal. But what is the expression for this size?

It is quite obvious that it is the capacity, as maximum production per year $Qty_{max}$, could be the most relevant measure of size. This capacity is easy to express for very specialized units, as found in oil industry (degasification, liquefaction, distillation and storage) or in petro-chemistry (ethylene and polypropylene). But many manufacturing units have various productions: a shipyard produces ships of different sizes and different uses, a car engines factory produces different engines of different power at different rates of production.

A solution sometimes used is to use a characteristic of the produced units, the tonnage of the ships, for example for a shipyard. It is a rather good solution, as long as basic technology (weld steel) remains the same one, and that the produced units remain roughly homothetic (cargo liners). But can one express with the ton a volume of production which would be composed for half (in tons) of cargo liners and passenger liners? There is certainly a higher content of work completed in a ton of passenger liner than in a cargo liner.

A solution would be then to use a characteristic of the workshops of production, the labor hours, for example. It is not a good solution, because if something is to be measured, it is the physical production and not the factors necessary to this production. For example the overall length of welding beads, and not quantity of hours of welders, because those can be more or less productive, well used and replaced by automatisms.

Indeed, this solution is practicable when the production units have a very homogeneous activity. It is the homogeneous sections method used in cost control. In these sections, the activity (which will be used as the divider of the budget in order to give the section cost rate) is expressed in units of work, characteristics of the activity (length of welding, surface treated, etc.). If the whole manufacturing unit has a homogeneous activity, then it is possible to measure its production with this kind of work unit.

The solution generally applied is to define units of reference, or **equivalent units**. It will be said that a ship methane tanker is equivalent to 1.3 bulk carrier of the same displacement. One will assimilate a top-of-the-range car to 1.5 bottom-of-the-range car.

This solution is criticizable: actually, explicitly or implicitly, the expert who defines the coefficients of assimilation will make it in proportion of production costs of the ships or cars. When thus one wants to express costs according to the production, one risks in fact to express, at least partially, costs according to costs. There is a risk of circularity.

Finally, nevertheless, this empirical solution is often the only applicable one.

## Cost of a Production

The power law (and the Chilton rule) gives a simple mean to extrapolate the cost of production when the size of the plant increases or decreases. The application of this law requires that the following conditions are met:

- the observed cases have similar natures of activity,
- the observed cases are located in a comparable economic context,
- especially, the observed cases are at equivalent load factors, or were brought back there (this is explained in Section 9.4.),
- the extrapolation will not exceed the extent of the reference case of a factor higher than 3,
- one has a reference frame made up of a sufficient number of observed cases.

If these conditions are met, the production cost $Cst$ in a plant of capacity $Qty_{max/yr}$ results from the cost $Cst_0$ of a reference plant of capacity $Qty_{0,max/yr}$ by:

$$\frac{Cst}{Cst_0} = \left( \frac{Qty_{max/yr}}{Qty_{0,max/yr}} \right)^s$$

$Cst_0$, $Qty_0$ and $s$ being beforehand given by the analysis of the reference frame.

It is possible to observe manufacturing units of different sizes using rigorously the same technologies, the same levels of automation, the same definition of positions.

This is however not observed on very important extents of size, about three only, usually. If we increase the sizes, automations will appear; for example in a foundry, an automatic station of weighing upstream will replace the manual station, and downstream robots of stripping will replace the men at this very hard station.

If we still increase the sizes, it is the basic technology which will change. The electric furnace will replace the cupola furnace, for example, in the case of iron foundries. It is more expensive, but much more productive.

What can we infer from this about the validity from the Chilton's law on a wide extent of manufacturing capacities?

1. Continuity is generally observed, simply because there is a crossing of the $Cst = f(Qty)$ curves, delimiting the fields of validity of various technologies.
2. Derivability is not certain. The curve typically shows breaks.
3. Each one of these individual curves follows in theory a power law.
4. It is not rare to be able to approximate the whole of the curves by a global power law.

## 9.3.2 Cost of the Products as a Function of the Size of Their Manufacturing Unit

We turn now to the cost of one item or $Cst_u$ and examine how this cost is going to change when the capacity of the plant changes:

$$Cst_u = \frac{Cst_{production}}{Qty_{actual}}$$

*Basic Analysis, for a Given Plant Capacity*
The cost of an item breaks up classically into:

- *Direct purchases* which includes:
  - bought parts composing the product;
  - raw material component;
  - energy directly integrated into the matter of the product (fusion, electrolyze).
- *Added value* which includes:
  - direct workforce (making manual operations directly on the product);
  - indirect workforce (necessary to operate the production unit);
  - indirect purchases (necessary to operate the production unit);
  - depreciation;
  - overheads (financial expenses, taxes, margin).

## Changing the Plant Capacity

If one passes a product from a manufacturing unit of capacity 100 to a unit of capacity 200, working both to 80% of their capacity, for example, one will observe:

- No impact on:
  - *Direct purchases*, the change of size of the production unit does not involve the purchase of larger quantities (thus at better costs). It involves only that operations are gathered in one unit instead of many.
  - *Direct workforce costs (the "touch labor")*, if technologies are equivalent, one has in both cases the same automatization level. The content of workforce of the product thus remains the same. One can note that the large units are often automated more than the small ones, but it is not always true, small units being able to be very automated; the link is too fuzzy to be taken into account in an algorithm.
- A fall of:
  - *Depreciation*, if the plant cost (the investment) is $Invest_{plant}$, its lifetime of $N$ years and its annual capacity $Qty_{max/yr}$, depreciation per product is (we use here the simple, linear, depreciation):

$$Depr = \frac{Invest_{plant}}{N \times Qty_{max/yr}}$$

Then, if the capacity goes from $Qty_{0,max/yr}$ to $Qty_{max/yr}$ the lifespan $N$ being constant:

$$\frac{Depr}{Depr_0} = \frac{Invest_{plant}}{Invest_{0,plant}} \times \frac{Qty_{0,max,yr}}{Qty_{max,yr}}$$

and, for:

$$\frac{Cst_{plant}}{Cst_{0,plant}} = \left(\frac{Qty_{max/yr}}{Qty_{0,max/yr}}\right)^s$$

$$\frac{Depr}{Depr_0} = \left(\frac{Qty_{max/yr}}{Qty_{0,max/yr}}\right)^{s-1}$$

$s$ range being between 0 and 1, any rise of:

$$\frac{Qty_{max/yr}}{Qty_{0,max/yr}}$$

will involve a fall of:

$$\frac{Depr}{Depr_0}$$

If

$$\frac{Qty_{max/yr}}{Qty_{0,max/yr}} = 2$$

for instance, and if $s = 0.7$ then

$$\frac{Depr}{Depr_0} = 2^{-0.3} = 0.81.$$

- *Indirect purchases and labor expenses*, their behavior is of the same as depreciation. Actually, control or maintenance of a plant twice more important does not require twice more personnel. In the same way, energy consumption does not double (to operate the plant; however the direct transformation of the material, as the electrolyze of alumina, or melting of metal, requires quantities of energy directly proportional to the weight of material: they are to be accounted for with raw material component). The power law thus returns a good account of their evolution according to the capacity.

*A Conventional Behavior of the Overheads*
The overheads as financial costs, taxes or margin, may be expressed as an absolute figure or as a percentage of a basis as purchases or depreciation. According to the convention, they are then to be taken into account with the part of the cost dependent or independent of the capacity, in addition to their basis.

*As a Conclusion*

We can consider that, when a product passes from a plant with a capacity $Qty_{0,\text{max/yr}}$ to another one with a different capacity $Qty_{\text{max/yr}}$, while remaining at the same load and the same technology:

- the material part $Cst_{u/mat}$ and workforce part $Cst_{u/wf}$ of the unitary cost $Cst_u$ of this product, remain constant;
- the remaining part $Cst_{u/capacity\_effect}$ varies as described on Figure 9.3:

$$\frac{Cst_{u/capacity\_effect}}{Cst_{0,u/capacity\_effect}} = \left(\frac{Qty_{\text{max/yr}}}{Qty_{0,\text{max/yr}}}\right)^{s-1}$$

$s$ being the extrapolation exponent of the cost of the production unit. The components of $Cst_{u/capacity\_effect}$ are: depreciation, indirect purchases, indirect workforce costs, some overheads:

$$Cst_u = Cst_{u/mat} + Cst_{u/wf} + Cst_{0,u/capacity\_effect} \times \left(\frac{Qty_{\text{max/yr}}}{Qty_{0,\text{max/yr}}}\right)^{s-1}$$

The "capacity effect": cost of one item as a function of the plant capacity.

# Plant Capacity and Load

## 9.3.3 Example

Let us consider the production cost of an aluminum ram. The data are the following one per unit:

| Cost items | $ | Comments |
|---|---|---|
| 1. Raw aluminum | 7.0 | 1.5 lbs |
| 2. Fusion energy | 1.0 | 22 kWh/lb @ 0.03 $/kWh |
| 3. Indirect purchases (fluids, other energy) | 1.5 | |
| 4. Fusion machinery and building | 1.8 | depreciation on 10 years, 1000 ton/year |
| 5. Moulding machinery and building | 2.7 | depreciation on 10 years, 1000 ton/year |
| 6. Specific tooling | 8.0 | depreciation on 10 000 parts |
| 7. In direct work force | 10.0 | |
| 8. Unitary control | 4.0 | |
| 9. Administration | 5.0 | |
| Total | 41.0 | |

The components of the formula giving the "scale" effect are:

$$Cst_{u/mat} \text{ line 1 + line 2} = \$8$$
$$Cst_{u/wf} \text{ line 8} = \$4$$
$$Cst_{u/capacity\_effect} = \$29$$
$$Cst_u \text{ (Total)} = \$41$$

Let us compute now the impact of an increase of the capacity of 100%, assuming a Chilton's law with $s = 0.6$:

$$Cst_{u/capacity\_effect} = 29 \times 2^{-0.4} = 29 \times 0.758 = 22$$
$$Cst_u = 8 + 4 + 22 = 34$$

The unit cost drops from $41 to $34!

## 9.4 How Does the Load (of a Plant with a Stable Capacity) Influence the Production Cost?

### 9.4.1 What Is the Load?

A manufacturing unit (a plant or part of it) generally produces several different items, each item being produced at a given quantity. The load is related to all the items; in order to quantify this load we need to "normalize" the items. This normalization process was described in Section 9.1.5 of this chapter. Once the items are normalized we can add their production quantities and say: the unit produced during the year a quantity of "equivalent items" named $Qty_{yr}$, whereas it was able to practically produce $Qty_{pr\_max/yr}$ (always equivalent items).

The load is then defined as:

$$\text{Load} = \frac{Qty_{yr}}{Qty_{pr.\_max/yr}}$$

It is important to understand that the load is relative to the equivalent items, irrespective of a particular item: the number of this particular item produced per year may (when we need it, it will be symbolized by $Qty_{part./yr}$) remain fix, but the load may change due to the other items.

### 9.4.2 The Causes of Variation of the Production Cost According to the Load

Let us consider a manufacturing unit with a theoretical capacity $Qty_{th.\_max/yr}$ and examine its total cost of production for a year $Cst_{t/yr}$ according to the quantity really produced $Qty_{yr}$ (or to its production load, which is equivalent).

The total cost of production $Cst_{t/yr}$ is the cost of the production of the whole period, the year, as used here; this cost is obviously a function of the quantity produced $Cst_{t/yr}(Qty_{yr})$ as illustrated on Figure 9.4. Let us locate on this figure three characteristic values of production:

1. production equal to nominal (the production which the unit must be able to hold continuously under good conditions, and for which it was normally conceived);
2. production equal to the theoretical maximum potential;
3. null production.

*Increasing Production*
When we leave the nominal production and let the production quantity increase, factors of direct "over-cost" appear gradually:

- overtime is introduced;
- the number of shifts is increased;

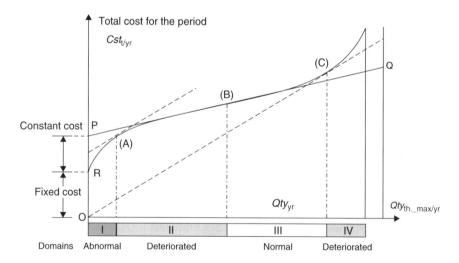

**Figure 9.4** Total production cost for a period.

- Saturday work is introduced, then Sunday work, then public holiday work;
- there is more and more recourse to the interim, etc.

Phenomena of indirect over-cost appear at the same time:

- maintenance must be made more quickly, with more means, by night;
- logistics appeals to more expensive means to maintain the supply safety, which becomes very critic;
- the supplies with subscription contracts (energy, fluids) are provided to raised rate;
- the organization is degraded, the rate of retouches increases.

In this field, the total cost increases more quickly than the production.

*Decreasing Production*
Let us set back to the nominal capacity, and now let the production quantity decrease; we will observe, beginning at certain production (point B), others factors of over-cost:

- short time working is introduced;
- times of changing tools and production starting, rejects and purges of batch beginning, remain constant, whereas the size of the batches decreases;
- the subscription of the supplies like energy and the fluids remains constant, whereas consumption decreases;
- the workforce is decreased, therefore the remaining personnel must hold more tasks, from where displacements from station to station, and errors which generate retouches;
- logistics calls upon more expensive means, to ensure a delivery which remains regular with smaller batches, or called upon a longer storage, therefore expensive.

In this field, the total cost decreases less quickly than the production.

### 9.4.3 The Total Production Cost as a Function of the Quantity $Cst_{t/yr}(Qty_{yr})$

*Remarkable Points*

*1 Point of Minimum Marginal Cost (Point B)*
One calls marginal cost $Cst_m(Qty_{yr})$ the derivative:

$$\frac{\partial Cst_{t/yr}(Qty_{yr})}{\partial Qty_{yr}}$$

of the total cost in respect to the production quantity during the period; it is also a function of the quantity produced during the period. This marginal cost is the change of the total cost when producing one more (equivalent) item.

There is a value of the production quantity – corresponding to the production quantity of point (B) – for which this derivative is minimum; the slope of the total cost is then minimum: this corresponds to line PQ of Figure 9.10. Geometrically, $Cst_m(B)$ is the point of inflection of the curve $Cst_{t/yr}(Qty_{yr})$. Over and under production corresponding to this point (B), one notes the over-costs mentioned above: points (A) and (C) have greater marginal costs.

## 2 Point of Minimum Average Cost: Nominal Capacity (Point C)

We turn now to the cost of one particular (equivalent) item. For a given $Qty_{yr}$ the total production cost is given by one point on the curve of Figure 9.4. The cost of one item or "unitary cost", $Cst_u$, is geometrically given by the tangent of the angle between the line joining this point and the point O and the horizontal axis. Arithmetically it is given by the ratio:

$$\frac{Cst_{t/yr}(Qty_{yr})}{Qty_{yr}} = Cst_u$$

This ratio is also a function of the quantity produced during the period.

This figure makes immediately appear that point (C) is particular: at point C, the unitary cost, which is minimal, is equal to the marginal cost: $Cst_m(Qty_{C/yr}) = Cst_u(Qty_{C/yr})$. It is for this production quantity that the unit is the most profitable, and, logically, it is for this production that it was designed. This production quantity $Qty_{C/yr}$ is thus the "nominal capacity" of the unit.

## 3 Point of Null Production

Is the total cost there null? No, because, although the unit is stopped, expenses do remain:

- On one hand, expenses of guarding and of minimal maintenance, insurance, and possibly certain taxes, subscriptions of certain adductions (water, power).
- On the other hand, the depreciation of the assets (buildings, machinery). We will call this cost $Cst_0$ the "fixed" cost.

## 4 Point of Maximum Production

It is a rather conventional concept. In fact, the theoretical maximum capacity $Qty_{th.\_max/yr}$ is not reachable. It is the practical maximum capacity $Qty_{max/yr}$ which marks the end of the total cost curve $Cst_{t/yr}(Qty_{yr})$. Unfortunately only $Qty_{th.\_max/yr}$ can be set without ambiguity. The difference between the two maxima varies with experts' advices, and motivation of operators ... One can always do better!

## 5 Point A

It is the point such as $Qty_A < Qty_B$ and such as the marginal cost has the same value as at point C:

$$\frac{\partial Cst_{t/yr}(A)}{\partial Qty_A} = \frac{\partial Cst_{t/yr}(C)}{\partial Qty_C}$$

## 6 Domains of Production

One can now define three domains of production:

1. a domain ranging between 0 and $Qty_A$ where the mode of production is abnormal;
2. two domains between $Qty_A$ and $Qty_B$ on one hand, beyond $Qty_C$ on the other hand, where the mode of production is deteriorated, but not abnormal;
3. a zone between $Qty_B$ and $Qty_C$ where the mode of production is normal.

## Constant and Proportional Charges for the Total Production Cost

To simplify the form of $Cst_{t/yr}(Qty_{yr})$, one can observe on Figure 9.4 that in the vicinity of (B), the change of the total production cost as a function of the quantity produced is very close to be linear. There is then a natural tendency[4] to return to a linear form that we write that in the vicinity of this point:

$$Cst_{t/yr} = Cst_{constant} + Qty \times Cst_{proportional}$$

This equation defines a $Cst_{constant}$ (corresponding to point P of Figure 9.4) and a $Cst_{proportional}$ corresponding to the slope of line PQ, which is the marginal cost at point B. Both costs are only "locally stable". They are always rather conventional; given the marginal cost at point (B) or $Cst_{proportional}$ $Cst_{constant}$ is deduced from it.

Mathematically this linear approximation of the total cost in the vicinity of point (B) is easily explained by the Taylor polynomial:

$$Cst_{t/yr}(Qty_{yr}) = Cst_{t/yr}(B) + \frac{\partial Cst_{t/yr}(B)}{\partial Qty_{yr}}(Qty_{yr} - Qty_B)$$
$$+ \frac{1}{3!}\frac{\partial^3 Cst_{t/yr}(B)}{\partial Qty_{yr}^3}(Qty_{yr} - Qty_B)^3 +$$

where the second order term does not appear because point (B) is an inflection point. One can easily show that if we neglect the terms of ranks equal or higher than 3 – which is reasonable assumption in the vicinity of (B) – then:

$$Cst_{constant} = Cst_{t/yr}(B) - \frac{\partial Cst_{t/yr}(B)}{\partial Qty_{yr}} Qty_B$$

$$Cst_{proportional} = \frac{\partial Cst_{t/yr}(B)}{\partial Qty_{yr}} = Cst_m(B)$$

Constant and proportional costs can be studied at any point defined by $Qty_{yr}$, using the same formulae; of course the values will be different because these costs are only locally stable. Whatever the point of interest, one can fix limits to the conventional values:

- $Cst_{proportional}(Qty_{yr})$ cannot be smaller than the minimum marginal cost observed at point B, and larger than:

$$\frac{Cst_{t/yr}(Qty_{pr.\_max}) - Cst_{t/yr}(0)}{Qty_{t/yr}}$$

- $Cst_{constant}(Qty_{yr})$ cannot be larger than OP.

---

[4] distinguishing between "fix" and "variable" costs, at a given production level, is very frequent in the literature. The cost we call "constant" is called "fix" in the literature, the cost we call "proportional" being call "variable".

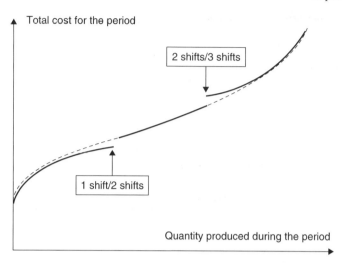

**Figure 9.5** The thresholds.

*Continuity of the $Cst_{t/yr}(Qty_{yr})$ Curve: Thresholds*

To be able to describe $Cst_{t/yr}(Qty_{yr})$ mathematically, we made an assumption of continuity which is seldom verified in reality.

To observe such a continuity, it would be necessary that the manufacturing unit should be made up of a great number of independent workshops (but although all affected by the variation of produced quantity). The causes of over-costs and the decisions of management would occur then in sequence and the $Cst_{t/yr}(Qty_{yr})$ curve would be about continuous.

It is not the general case. When a factory introduces a third shift, it is often on the whole plant, or on the major part of it, simultaneously (Figure 9.5).

The $Cst_{t/yr}(Qty_{yr})$ curve is then discontinuous. So that, if a small variation of the production quantity requires a third shift, we shall notice a break of the curve for this quantity.

It is one of the reasons which make it very difficult to observe practically the shape of the curve of Figure 9.4. It is nevertheless essential to set this modelization if one wants to normalize the unitary costs $Cst_u(Qty_{yr})$ as a function of the production quantities, without going into detail (unavailable generally) of the $Cst_{t/yr}(Qty_{yr})$ graph.

*Unitary and Marginal Costs $Cst_u(Qty_{yr})$ and $Cst_m(Qty_{yr})$*

Figures 9.6–9.10 show the impact of the size of the plant and its load on the unitary costs.

Figure 9.6 shows the total production cost for the period (full line), the marginal cost, its derivative (dotted line) and the unitary cost or mean cost (mixed line)
Notice that:

- minimum of marginal cost occurs earlier than minimum of unitary cost when $Qty$ increases,
- minimum of marginal cost is lower than minimum of unitary cost,

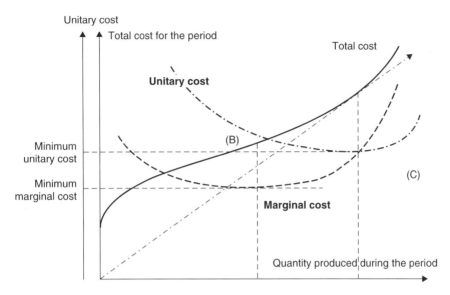

**Figure 9.6** Unitary and marginal costs as a function of the produced quantity for a given plant capacity.

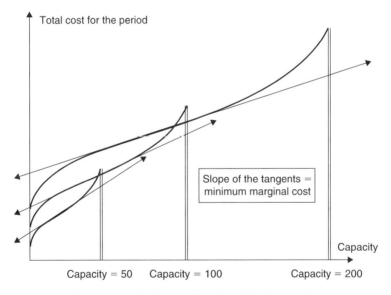

**Figure 9.7** Marginal cost as a function of the plant capacity.

- if unitary cost is minimum, it is equal to marginal cost,
- for $Qty = 0$, marginal cost is finite, unitary cost is infinite.

In Figures 9.7 and 9.8, for a given produced quantity, the marginal cost is the slope of the tangent to the total production cost at this point and the unitary cost the slope of the secant coming from the origin. When the capacity increases, the minimum marginal cost and the minimum unitary cost decreases.

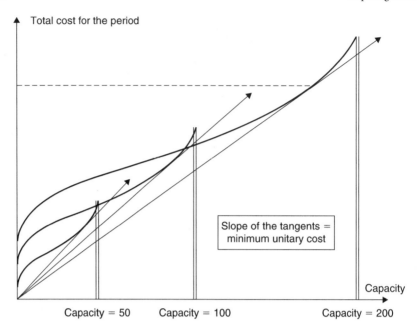

**Figure 9.8** Marginal cost as a function of the plant capacity.

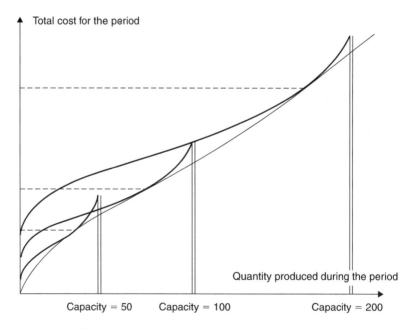

**Figure 9.9** Quantities which give the minimum unitary costs.

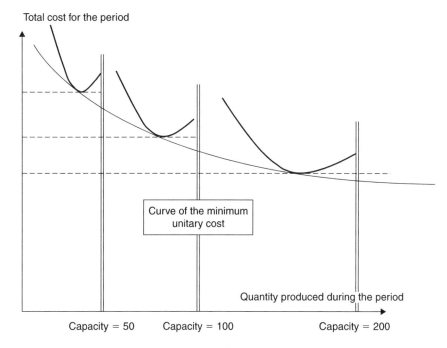

**Figure 9.10** Quantities which give the minimum unitary costs.

In Figures 9.9 and 9.10, when the capacity increases, the quantities which give the minima of the unitary cost can be displayed on the total cost curve or on the unitary costs.

### 9.4.4 Unitary Costs According to the Load of Plant

The purpose of this section is to study how $Cst_u$, the cost of one particular item, changes when the load of the manufacturing unit changes. For this purpose we will break down this cost into three parts:

1. A part which is independent on the load, but is proportional to the quantity of this particular item.
2. A part which is sensitive to the load, called $Cst_{u/load\_effect}$.
3. A part which is independent on the load, but decreases with the quantity of this particular item.

*Breaking Down the Cost According to the Sensitivity to the Load*

How does the full production cost $Cst_{t/yr}$ change with the load?

Let us seek, among the various components of the total production cost, those which are sensitive to the load and those which are not sensitive to it.

## Direct Purchases

One recalls that the direct purchases are those which remain in the product: two cases are to be considered:

1. *Parts*: They are components specific to the production of the unit. They are produced by suppliers whose load of the manufacturing units are sensitive to the quantity of this particular item. One assimilates as parts the especially developed materials (paintings, specific coatings, etc.).

   The bought specific parts are treated like the parts produced by the studied manufacturing unit. The reasoning exposed in this chapter can thus reproduce for each subset, by analytically breaking up the product according to its nomenclature. It is interesting and possible only for the important subsets. For the other ones, one will arbitrarily decide to assimilate them to $Cst_{u/load\_effect}$ which is the part of the cost excluding materials. In this discussion, one supposes that no analytical decomposition was made.

2. *Materials*: They are the raw materials (Steel, aluminum, polymers, etc.). In general these materials are standard and are the subject of a world market. In general the quantities supplied by the manufacturing unit are low before the market and will be without influence on its level. One assimilates to the materials the standard parts (fixings, parts bought on catalogue).

   Thus, material share of the unit cost: $Cst_{u/mat}$ = constant; the material part of the item cost does not change with the production load. (NB: the material share of the total cost $Cst_{t/yr}$ is thus variable. It is a share of the variable expenses.)

## Specific Tooling

Let us return to the aluminum moulding installation case. A given item can have required a very complex and expensive mould and this item may have not met success. Nevertheless, the workshop or the factory can be normally loaded. The curve $Cst_{t/yr} = f(Qty_{yr})$ will not take in account the phenomenon if $Qty_{yr}$ represents only the global output of the workshop.

To give an account of it, it is necessary to define a share specific tooling cost *Tool* which is composed of the sum $\Sigma_i \, Tool_i$ of the tools specific to each item $i$. One could consider that *Tool* includes, as $Cst_{t/yr}$, depreciation and expenditures (maintenance, handling) variable with the production quantity $Qty_{yr}$.

In practice one considers that *Tool* is made up only of depreciation. Therefore this cost is independent of $Qty_{yr}$:

$$\text{Tool} = \text{constant}$$

NB: the share tooling in the unit cost $Cst_{u/tool}$ is thus variable with the production quantity.

## Direct Workforce and Tooling Shares: Comparison with the Cost Capacity Relation

The direct workforce is the one which has a direct contact with the product ("touch labor"). In the automated installations it is limited to the operations of loading and unloading the lines, and the final controls.

The approach is similar to that exposed to the Section 9.3.2 for the capacity. But take care, the composition of $Cst_{u/capacity\_effect}$ and $Cst_{u/load\_effect}$ is not exactly the same: $Cst_{u/capacity\_effect}$ includes $Cst_{u/tool}$ and excludes the touch labor $Cst_{u/wf}$, the opposite being true for $Cst_{u/load\_effect}$.

Indeed, a variation of capacity of the installation leaves unchanged the times and the rates of direct labor. On the contrary a variation of load will have an influence on the rates of labor (night-work, holiday-work).

Conversely a variation of capacity of the installation causes in general a variation, in the long term, of the capacity of the specific tools (e.g. number of cavities of the moulds). But a variation of the load factor of the factory does not have any influence on the specific load factor of the tool of a particular part.

## Variation of the Elements of the Unitary Cost According to the Load

The general relationship is given by:

$$Cst_u = \underbrace{Cst_{u/mat} + Cst_{u/load\_effect}}_{\text{manufacturing cost}} + \underbrace{\frac{\text{Specific\_tooling\_total\_cost}}{N \times Qty_{part./yr}}}_{\text{tooling depreciation cost}}$$

### The "load effect": cost of one item as a function of the production load

| | | |
|---|---|---|
| $Cst_{u/mat}$ | raw material share | Constant |
| $Cst_{u/load\_effect}$ | share of the cost variable with the production of the shop | Variable |
| Particular tooling total cost | – | Constant |
| $Qty_{part./yr}$ | annual production of the particular item | |
| $N$ | years of production of this item | |

### Variation of $Cst_{u/load\_effect}$ According to the Load

We saw in Section 9.2.1 the law of variation of the costs $Cst_{t/yr}$ and $Cst_u$ for the whole production.

The cost $Cst_{u/load\_effect}$ of a part corresponds to the cost $Cst_u$ of a factory. (NB: "Correspond", and not "is equal"; as the cost $Cst_{u/load\_effect}$ of a part often covers in fact the costs of several factories which we suppose, by simplification, that they have the same load and the same variation of the cost according to the load.)

The description of the law of variation of $Cst_{t/yr} = f(Qty_{yr})$ in a factory could be done by a non-linear equation (see Chapter 3.8). We will not follow this way here. Indeed, since decades, the financial controllers of the factories are accustomed to build the budgets of the $N+1$ year of their plants starting from the examination of the fixed and variable shares of the budget of year $N$. We can make profit of their observations:

- the fixed/variable proportions of the budget are rather comparable from one factory to another one in the same technology, and for equivalent capacities.
- the fixed/variable proportions of the budget are not constant with the load.

### An Example

In this example, we are interested in what happens to the unitary cost of an item when the quantity produced for this particular item $Qty_{part./yr}$ goes from 0 (the factory does not produce this item) to $Qty_{yr}$ (the factory only produces this item); this range is divided into 10 intervals. This capacity $Qty_{yr}$ remains constant and equal to 1800 (equivalent) items.

**Table 9.2** Simulation of the production of the same item at different quantities.

| 1 | 0 | A | | Deteriorated | | B | Normal | C | Q | Deteriorated |
|---|---|---|---|---|---|---|---|---|---|---|
| 2 | Abnormal | | | | | | | | | |
| 3 | 1 | 2 | 3 | 4 | 5 | 6 | 7 | 8 | 9 | 10 |
| 4 | 1800 | 1800 | 1800 | 1800 | 1800 | 1800 | 1800 | 1800 | 1800 | 1800 |
| 5 | 0 | 200 | 400 | 600 | 800 | 1000 | 1200 | 1440 | 1600 | 1800 |
| 6 | 20 | 360 | 560 | 750 | 916 | 1070 | 1224 | 1440 | 1728 | 2250 |
| 7 | | 1.8 | 1.4 | 1.25 | 1.145 | 1.07 | 1.02 | 1 | 1.08 | 1.25 |
| 8 | 0.00 | 0.11 | 0.22 | 0.33 | 0.44 | 0.56 | 0.67 | 0.80 | 0.89 | 1.00 |
| 9 | | −200 | −200 | −200 | −200 | −200 | −200 | −240 | −160 | −200 |
| 10 | 200 | 200 | 200 | 200 | 200 | 200 | 240 | 160 | 200 | |
| 11 | | −340 | −200 | −190 | −166 | −154 | −154 | −216 | −288 | −522 |
| 12 | | 200 | 190 | 166 | 154 | 154 | 216 | 288 | 522 | |
| 13 | | 1.7 | 1 | 0.95 | 0.83 | 0.77 | 0.77 | 0.9 | 1.8 | 2.61 |
| 14 | | 1 | 0.95 | 0.83 | 0.77 | 0.77 | 0.9 | 1.8 | 2.61 | |
| 15 | | 20 | 360 | 560 | 750 | 916 | 1070 | 1224 | 1440 | |
| 16 | | 560 | 750 | 916 | 1070 | 1224 | 1440 | 1728 | 2250 | |
| 17 | | | 1.80 | 1.40 | 1.25 | 1.15 | 1.07 | 1.02 | 1.00 | |
| 18 | | | 1.25 | 1.15 | 1.07 | 1.02 | 1.00 | 1.08 | 1.25 | |

### 1. Simulation of a smooth total production cost

Table 9.2 presents the budget $Cst_{t/yr}$ (line 6) and the unitary cost $Cst_u$ (line 7) of a production for various quantities $Qty_{part./yr}$ (line 5) and the same capacity $Qty_{yr}$ (line 4). The load is:

$$\frac{Qty_{part./yr}}{Qty_{yr}}$$

This table must be read the following way:

Line 1    Remarkable points
Line 2    Production domain
Line 3    Interval $i$ (from 1 to 10) of the load
Line 4    Plant capacity
Line 5    Quantity produced of the studied item
Line 6    Total production cost (k$)
Line 7    Unitary cost (k$): line 6/line 5
Line 8    Plant load: line 5/line 4
Line 9    Change in quantity from interval $i - 1$ to $i$: $Qty_{i-1} - Qty_i$
Line 10    Change in quantity from interval $i + 1$ to $i$: $Qty_{i+1} - Qty_i$
Line 11    $Cst_{t/i-1} - Cst_{t/i}$ change of the total cost
Line 12    $Cst_{t/i+1} - Cst_{t/i}$ change of the total cost
Line 13    $K_i = (Cst_{i+1} - Cst_i)/(Pns_{i+1} - Pns_i)$ (Adjustment with the rise)
Line 14    $K'_i = (Cst_{i-1} - Cst_i)/(Pns_{i-1} - Pns_i)$ (Adjustment with the fall)
Line 15    Total cost $Cst_{t/i-1}$ = line 6 + line 9 × line 13
Line 16    Total cost $Cst_{t/i+1}$ = line 6 + line 10 × line 14
Line 17    Unit cost $Cst_{u/i-1}$ = line 15/(line 5 + line 9)
Line 18    Unit cost $Cst_{u/i+1}$ = line 16/(line 5 + line 10)

The financial controllers compute the coefficients of adjustment of the budget $K$ and $K'$ corresponding to a rise and to a fall, with the property: $K_i = K'_{i+1}$. The curve

# Plant Capacity and Load

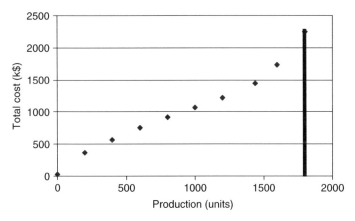

**Figure 9.11** Simulation of a smooth cost curve.

**Table 9.3** Simulation of a cost curve with thresholds.

| 1 | | | | B | | | C | | Q | |
|---|---|---|---|---|---|---|---|---|---|---|
| 2 | 1 shift | | | 2 shifts | | | | | 3 shifts | |
| 3 | | 4 | 5 | 5.1 | 6 | 7 | 8 | 9 | 9.1 | 10 |
| 4 | 1800 | 1800 | 1800 | 1800 | 1800 | 1800 | 1800 | 1800 | 1800 | 1800 |
| 5 | 400 | 600 | 800 | 801 | 1000 | 1200 | 1440 | 1600 | 1601 | 1800 |
| 6 | 560 | 680 | 820 | 950 | 1070 | 1224 | 1440 | 1650 | 1800 | 2250 |
| 7 | | 1.133333 | 1.025 | 1.186017 | 1.07 | 1.02 | 1 | 1.03125 | 1.124297 | 1.25 |
| 8 | 0.33 | 0.44 | 0.45 | 0.56 | 0.67 | 0.80 | 0.89 | 0.89 | 1.00 | |
| 9 | −200 | −200 | −1 | −199 | −200 | −240 | −160 | −1 | −199 | |
| 10 | 200 | 1 | 199 | 200 | 240 | 160 | 1 | 199 | −1800 | |
| 11 | −120 | −140 | −130 | −120 | −154 | −216 | −210 | −150 | −450 | |
| 12 | 140 | 130 | 120 | 154 | 216 | 210 | 150 | 450 | −2250 | |
| 13 | 0.60 | 0.70 | 130.00 | 0.60 | 0.77 | 0.90 | 1.31 | 150.00 | 2.26 | |
| 14 | 0.70 | 130.00 | 0.60 | 0.77 | 0.90 | 1.31 | 150.00 | 2.26 | 1.25 | |

$Cst_t = f(Qty)$ can then be approximated by a sequence of segments of line of slope $K_i$ (Figure 9.11).

It is then always possible to find the $Cst_t$ on the right and on the left of each value starting from $K$, and the cost $Cst_u$ (Table 9.2).

### 2. Simulation of a total production cost with thresholds

The same technique makes it possible the description of thresholds: see Table 9.3 and Figure 9.12.

The table must be read the following way:

Line 1:     Remarkable points
Line 2:     Number of shifts and position of the thresholds
Line 3:     Interval $i$ (from 1 to 10) of the load
Line 4:     Plant capacity
Line 5:     Quantity produced of the studied item
Line 6:     Total production cost (k$)
Line 7:     Unitary cost (k$): line 6/line 5
Line 8:     Plant load: line 5/line 4

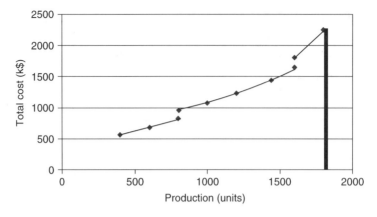

**Figure 9.12** Simulation of a smooth cost curve.

Line 9:     Change in quantity from interval $i-1$ to $i$: $Qty_{i-1} - Qty_i$
Line 10:    Change in quantity from interval $i+1$ to $i$: $Qty_{i+1} - Qty_i$
Line 11:    $Cst_{t/i-1} - Cst_{t/i}$ change of the total cost
Line 12:    $Cst_{t/i+1} - Cst_{t/i}$ change of the total cost
Line 13:    $K_i = (Cst_{i+1} - Cst_i)/(Pns_{i+1} - Pns_i)$ (Adjustment with the rise)
Line 14:    $K'_i = (Cst_{i-1} - Cst_i)/(Pns_{i-1} - Pns_i)$ (Adjustment with the fall)

### 3. Solving the problem

Let us return to our initial problem, which is to find a law making possible to go from a value $Cst_u$ with a given load to the value $Cst_{u/ref}$ that the item cost would have for a reference load.

We consider here only the factory costs. Notice thus that the costs $Cst_t$ and $Cst_u$ are here made up only of added value and subject to variation with the load. We shall then use the notation $Cst_{u/load\_effect}$ as shown in the general relation above.

We let apart the interval the abnormal zone located between points 0 and A. In this zone, the problem is not to describe what occurs, but to leave it as quickly as possible! The required reference data necessary to this description are in general missing, or are of doubtful validity.

In the other zones, we see that it is possible to go from any value $Cst_u$ with a given *Load* towards a load of reference (here 80%) if we have a table giving the coefficient of $K_i$ adjustment for each one of the $i$ sections of load.

The relation making possible to do this is:

- with the rise:

$$Cst_{u/load\_effect} = Cst_{u/i-1} \times Load_{i-1} + K_{i-1} \times (Load_{i-1} - Load_i)/Load_i$$

- with the fall:

$$Cst_{u/mat} = \frac{80000}{12000} = \$6.6$$

It is thus possible to pass $Cst_{u/load\_effect}$, section to section, from the load observed with the reference load (Table 9.4).

Plant Capacity and Load

**Table 9.4** Going from one cell to another one.

| 1 | 0 | A | | | | B | | C | | Q |
|---|---|---|---|---|---|---|---|---|---|---|
| 2 | | Abnormal | | Deteriorated | | | Normal | | | Deteriorated |
| 3 | 1 | 2 | 3 | 4 | 5 | 6 | 7 | 8 | 9 | 10 |
| 4 | | | 0.22 | 0.33 | 0.44 | 0.56 | 0.67 | 0.80 | 0.89 | 1 |
| 5 | | | 0.95 | 0.83 | 0.77 | 0.77 | 0.90 | 1.80 | 2.61 | |
| 6 | | | 1.40 | 1.25 | 1.15 | 1.07 | 1.02 | 1.00 | | |
| 7 | | | | | | | | ↓ | | |
| 8 | | | | | | | | 1.00 | 1.08 | 1.25 |
| 9 | | | | | | | | ← | | |

**Table 9.5** Unit cost data for the aluminum ram.

| Cost items | $ | Comments |
|---|---|---|
| 1. Raw aluminum | 7.0 | 1.5 lbs |
| 2. Fusion energy | 1.0 | 22 kWhr/lb @ 0.03 $/kWhr |
| 3. Indirect purchases (fluids, other energy) | 1.5 | |
| 4. Fusion machinery and building | 1.8 | depreciation on 10 years, 1000 ton/yr |
| 5. Moulding machinery and building | 2.7 | depreciation on 10 years, 1000 ton/yr |
| 6. Specific tooling | 8.0 | depreciation on 10 000 parts |
| 7. In direct work force | 10.0 | |
| 8. Unitary control | 4.0 | |
| 9. Administration | 5.0 | |
| Total | 41.0 | |

The table must be read the following way (row 8 is the reference load):

Line 1: Remarkable points
Line 2: Production intervals
Line 3: Interval $i$ (from 1 to 10) of the load
Line 4: Load
Line 5: K
Line 6: $Cst_{u/load\_effect}$ for a production rise
Line 8: $Cst_{u/load\_effect}$ for a production fall

*Numerical Application*

Let us return to the production cost of the aluminum ram as introduced in the Section 9.3.3. The cost data per unit reproduced here (Table 9.5).

The question we have to answer to is: what is the impact of an increase of the load from 56% to 80% (of equivalent items) and a rise of 20% of the quantity to be manufactured for this ram?

The components of the formula giving the "load effect" are:

$Cst_{u/mat}$ line 1 + line 2 = $8
$Cst_{u/tool}$ line 6 = $8      Total cost of tooling = $80 000
$Cst_{u/load\_effect}$ = $25
$Cst_u$ (Total) = $41

**1. About the Increase of the Load**

What is the impact of a rise of the load from 56% to 80%? We can assimilate this shop to the one described in Table 9.4. Therefore:

$$Cst_{u/load\_effect} = 25 \times 1/1.07 = \$23.4$$

instead of $25 for a load of 56%.

## 2. About the Increase of the Quantity to be Manufactured

The impact of an increase of the quantity affects only the depreciation of the tooling for this item. As the total cost of tooling is $80 000, the depreciation per unit is now given by:

$$Cst_{u/mat} = \frac{80000}{12000} = \$6.6$$

instead of $80\,000/10\,000 = \$8$

## 3. Synthesis

As $Cst_{u/mat}$ remains constant, then the unit cost becomes $8 + 6.6 + 23.4 = \$38$ instead of $\$41$.

## 9.5 In Practice

### 9.5.1 The Different Points of View

We can different points of view, with more or less accessibility to the data:

- Purchaser
- Estimator
- Financial controller

Each one of these different persons needs modeling for different reasons.

*The Point of View of the Procurement Officer*

In this first case, one works on prices, and not on costs. Nevertheless in the negotiation of the arguments of cost are used. In his situation the procurement officer must know:

- $Tool_{part./t}$, the investment made by the supplier for this particular item,
- $Qty_{part./pr.yr}$ the practical capacity of this tooling,
- $Qty_{part.}$, effective production quantity, corresponding to his/her orders.

In addition, he/she can estimate $Cst_u$, according to the information given at the time of the negotiation. In simple cases (casting of aluminum, plastic moulding) he/she can also estimate $Cst_{u/mat}$, the material share.

But he/she does not know, in theory:

- the characteristics of the plant: its capacity $Qty_{max/time}$, its production quantity $Qty_{yr}$ (of equivalent items), and its load;
- the parameters of the curve $Cst_{t/yr} = f(Qty_{yr})$ of this plant.

In practice, the purchaser can at least consider $Cst_{u/load\_effect}$ constant and calculate the impact of the variation of material prices on $Cst_{u/mat}$ and the impact of the volume of order on the tooling depreciation for this item

$$\frac{Tool_{part./t}}{Qty_{part.}}$$

Plant Capacity and Load

*The Point of View of the Management Controller*
Here, we speak about cost and not about price. We know $Qty_{yr}$, $Qty_{max/time}$ and the load of the plant.

The parameters of the curve $Cst_{t/yr} = f(Qty_{yr})$ are known in detail. In current management practice, one does not make any modelization of this curve; one knows for each interval the quantities of units of work (time) and the rate of the sections for each year. One can then recompute analytically the rates of sections permanently, for all the budgetary assumptions, and the costs which derive from it.

In practice, it is long and difficult to make analytical computations. However at the beginning of project, it is necessary to make many approaches with a precision rather weak, and few entries.

The financial controller makes the modelizations described above. They enable him/her to treat very quickly the estimates in early phases. The management controller has all the data he/she needs to refine these modelizations as much as wished, and only the cost benefit ratio of his/her work will determine its complexity.

*The Point of View of the Cost Estimator*
The estimators are in an intermediate situation.

Compared to the purchasers, the estimators are often elected to carry out audits near the suppliers. They can then appreciate the capacities, production and load of their factories.

Compared to the management controllers, they have much more budgetary simulations to carry out. They start from values of $s$ (the exponent of the extrapolation formula) and $K$ (the coefficient of adjustment) measured by the management controllers, then carry out assimilations, from the workshops of their own company or those of the suppliers.

The cost estimators have then a specific mission: *building a reference frame of estimate*. For doing it, the collected raw data must be normalized by establishing them to the same contour and to the same general economic conditions (localization, inflation, training) and to the same conditions of produced quantity. This is the process of normalization.

On this last point, modelizations described above make it possible to carry out the operation of normalization on capacity and load.

At the time of the estimate, one will of course have to make the symmetrical operation, with the same model, to adapt the costs of the reference frame to the imposed conditions.

### 9.5.2 A Few Other Comments

*Significance of $Cst_{u/capacity\_effect}$ and $Cst_{u/load\_effect}$ in Relation to Time*
The functions $Cst_{u/capacity\_effect} = f(Qty_{yr})$ and $Cst_{u/load\_effect} = f(load)$ are not at all in the same order of time.

The reaction to the variations is immediate. The reaction to variations of capacity is sensitive only at the end of the construction of new units, or at the enlarging the old ones.

In a normal economic and political environment, the adaptation to the rise of the demand by the increase in the capacities is a certainty.

One can thus consider that the evolutions of $Cst_{u/capacity\_effect}$ and $Cst_{u/load\_effect}$ account for the same phenomenon, but on different scales of time.

### Bottlenecks

Without going up to the rebuilding of installations, the plant can adapt itself, sometimes rather quickly, to an increase of the demand. There are indeed bottlenecks of which the smallest determines the capacity. It is often possible to get rid of the most constraining necks with the help of a moderate investment.

The historical curves reporting the production costs, according to a production which grows, generally integrate these removals of bottlenecks. They then do not follow exactly a law in $Cst_{u/load\_effect}$: one generally does not observe a "pure" curve because the plants adapt themselves quickly.

### Load of a Factory and Loading of a Production Line

A factory which includes several lines can see one of its lines saturated and another one under loaded. The phenomena of over-cost described above are always observed, but are attenuated: one will not make layoff, but one will pass workforce from one line to the other one, for example. But the tensions on the final improvements and logistics will remain.

A study which would note only the links between the cost of the products of one of these lines and its load, without taking account of the total load of the factory, would not give a correct description of reality.

### Versatility and Reconditioning

The versatility (immediate adaptation to a change) consists in the capacity of a production line to accept different productions. The price to pay is an investment definitely more expensive.

The versatility can avoid the arrival of overloads. If such a line manufactures indifferently three products A, B, C, and if (with a constant total quantity) the distribution between these products changes, one will not have any overload whereas one would have had one if there had been dedicated lines.

The aptitude for the reconditioning (postponed adaptation to a change) is a mean to of avoid the over-costs imposed by the versatility. One over-invests at the time of the installation of the specific lines A, B, C, for example, so that each line can be transformed quickly and with a moderate investment, at the time of the annual stop for example, to manufacture one of the two other products. If for instance product A, due to lack of success, is stopped, one will be able to transform its production line to make product B, which meets success.

The statistics drawn from the results of units which do not have the same degree of versatility are thus likely to lead to results different enough for the influence of the load on the cost.

# 10 Other Normalizations

One can consider other normalizations, depending on one's activity. Let us mention for instance the normalization related to the plant localization and the normalization related to the position in the life cycle of a product.

## 10.1 Normalization Related to the Plant Localization

This type of normalization has been studied by the engineering companies working for the petro-chemical industries. These industries have to build plants in different locations in the world, these locations differing by the level of development, the availability of labor, the regulations, the tax policy, etc.

As it is very specific, the reader may consult Ref. [18], p. 223.

## 10.2 Normalization Related to the Position in the Life Cycle of a Product

A cost of a product changes as it progresses in its life cycle. A new product, using new technologies, is always expensive at the beginning of its life cycle. This comes from the fact that the technology is not mastered yet, and that, consequently the level of defective ones is high, plus the fact that quantity produced is very small.

Several factors have to be considered for describing the changes in the cost. Two concepts are important in this life cycle: the time it takes between the idea of a new product and its mass production on one hand, the decrease of the cost during this time on the other hand.

Let us illustrate the first one for well-known products. Let us start with the steam engine:

- 1690: Denis Papin (1647–1712) describes the atmospheric engine (so called because he believed that the atmospheric pressure was really the force which produces the work) using one cylinder and one piston.
- 1698: Thomas Savary (1651–1715) presents a patent for this type of engine, but without a piston (then it cannot be called an "atmospheric engine"): the steam enters in a cylinder which is cooled down. This causes a depression which could be used as a pump.

- 1712: Thomas Newcomen (1663–1729) creates an engine with a piston filled with steam at the atmospheric pressure (a counterweight equilibrates the weight of the piston) and then cooled by water. This generates a depression which actuates the piston. This machine was installed in many mines.
- 1769: James Watt (1736–1819) adds a separated condenser which allows to keep the cylinder at a constant temperature. In 1782, he adds a distribution of steam on both faces of the piston, creating a "double action" machine, four times more efficient than Newcomen's machine. This machine was really the basis of the "external combustion engine" which was used during the whole 19th century, with of course many improvements.

This was really the start. Then came the applications, the most famous being the locomotive:

- 1804: Richard Trevithick develops the first locomotive (the "New Castle").
- 1814: George Stephenson creates the first locomotive "Blücher".
- 1825: First commercial link from Stockton to Darlington (UK).
- 1829: Stephenson develops the machine with multi-pipes heater.
- 1830: The link Liverpool to Manchester is opened.
- 1902: The extra heater is invented by Wilhem Schmidt.

The locomotive powered by a steam engine saw it apogee in 1941 with the "Big Boy" of Union Pacific. During the 1950, diesel locomotives (invented in 1913 in Germany) started to replace the steam engine locomotive.

It is interesting from an historical point of view to establish a parallel with the development of the electricity:

- 1800: Alessandro Volta creates the battery.
- 1820: Hans Oersted discovers the principle of electromagnetism.
- 1821: Michael Faraday (1791–1867) develops the principle of the direct current engine.
- 1858: The electrical arc introduces the public lightning in the Kent.
- 1868: Georges Leclanché creates a commercial, reliable, battery.
- 1873: Zénobe Théophile Gramme creates the first electrical engine reliable enough.
- 1878: Joseph Swann (1828–1914) demonstrates the lightning by filament.
- 1879: Thomas Edison (1847–1931) creates his incandescence lamp.
- 1888: Nikola Tesla creates the electrical engine driven by alternative current.

What conclusions may the cost analyst draw from these stories (and many others which followed a similar pattern)? The first conclusion is that the development of the steam engine, from the first prototype to the really efficient machine, lasted about 80 years and that 70 years were needed for developing the first practical application of electricity. Between the discovery of a new "secret" of nature and its application, a lot of time is necessary, but this time shows a trend to a reduction.

However the development of the locomotive lasted no more than 30 years. Once the secret of nature is understood, developing a completely new application needs about one-third of the time. It is interesting to see that the development of the internal combustion engine – from Etienne Lenoir who patented the first engine using the natural gas and a spark-plug in 1860 to Albert de Dion and Georges Bouton manufacturing the light engine turning at 1500 turns/min in 1895 – also a time lag of 30 years was needed.

## Other Normalizations

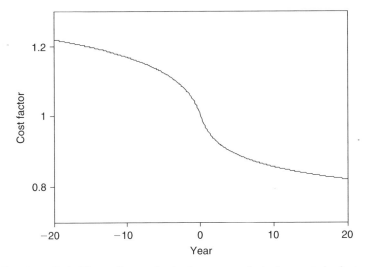

**Figure 10.1** Typical "natural" cost reduction for a new product using new technologies.

During all this time cost decreases rapidly at the beginning, due to the facts that the principles become well known, that machines were built to manufacture the products in a more efficient way and mainly that some people invest a lot of money for industrialization, the typical example being the electronic circuits.

As far as the cost is concerned, two different phenomena must be taken into account:

1. The "normal" cost reduction due to the fact that the manufacturing process is progressively mastered. This cost reduction is limited. A typical reduction is illustrated in Figure 10.1. The presence of an inflexion point is a characteristic of this curve: it comes from the fact that the cost decreases rapidly at the beginning, but cannot do that for ever ... The year corresponding to this inflexion point is called "the point of maturity"; it is generally – as it is on the figure – chosen as the reference point; this choice is not specially a good one because, if it is well defined from a mathematical point of view (the second derivative cancels itself), it is however poorly defined from a physical point of view (it is there that the decrease is the steepest). But there is no other particular point.
2. Physically there is a correlation between the position of this point and the data where the sales take off. Both are located at about the same year.
3. The reader should be aware of the fact about the slope at the inflexion point: the more advanced (the definition of a more "advanced" product is given in Chapter 9) a product, the steeper the slope.
4. The cost reduction due to the increased production capacity. This effect is developed in Chapter 9.

For normalization purpose, the inflexion point is generally selected. This means that the cost analyst should adjust the collected cost to the cost he/she would have seen if the production were made at this point.

# Part III
## About General Models

# Part Contents

Chapter 11  **Definition of a General Model: Example**
Chapter 12  **Building a General Model**
Chapter 13  **New Concepts in General Modelization for Hardware**
Chapter 14  **Modelization in Other Classes**
Chapter 15  **A Word About the Future**

A short word at the history of science helps illustrate the fundamental differences between specific and general models.

Any science first looks at the facts and records them: in our modern language, any science starts by building databases. In the domain of astronomy the fantastic amount of data accumulated by great observers, such as Copermic, Tycho Brahé, etc., deserves our admiration.

The second step – not necessarily carried out by the same persons – is to try to establish and quantify correlations between variables which, apparently, may seem different. Once a good correlation has been demonstrated between these variables – this involves what is called in Volume 2 "data analysis" – it is rather tempting a build a mathematical relationship between these variables; these relationships do not "explain" anything; they are just a tentative to group in a few equations what we know about the facts. The nice thing about them is that they make us able to predict values of one variable when the other one(s) is (are) known, as long as the previsionist remains in the same area (we would say, in the domain of cost, in the same "product family"). These relationships are called "laws" (we would say in the cost domain cost estimating relationships). There are plenty of such relationships in all sciences; just remember: the three Kepler laws, the Kirchoff laws, the Van der Wals law, the law of light emission by the black body, etc. The authors of these laws did not know the regression analysis and generally work by curve fitting, but the idea is the same.

The third step is far more recent; it implies to look "below the facts" in order to understand them (which means explaining by investigating things more in depth and finding reasons for their behavior).[1] This is done by finding abstract concepts (such as the forces in the description of motion, the fields in electrodynamism, the entropy in thermodynamics, the quanta in light emission, etc.) from which the facts could be "explained". The mathematical support then becomes a must, as it is the only way the human mind can work with abstract concepts. The great names (the "giants" to cite a word used by Newton)[2] in this respect are Newton, Maxwell, Boltzman, Planck, Einstein, etc. The set of equations they developed, generally a very limited set from which all phenomena can be predicted,[3] is generally called a "theory".

In the cost domain, the abstract concept that throws a powerful light on the cost behavior is the "product structure". This concept was described by Lucien Géminard in France and maybe others. This concept, which will be more developed in Chapter 13, helped create a general "theory" of cost behavior. But it is the only time I will use this term of "theory" in our domain and for three reasons:

1. The first reason is that human behavior is far less predictable than natural phenomena in the physical sciences. Therefore, the fantastic level of precision often attained in the physical sciences cannot be obtained in the domain of cost. The word "theory" in the

---

[1] It is well known that we never "understand" nature fully, by a step by step analysis requiring less and less hypothesis: understanding nature really means reducing the number of the hypotheses which are necessary for predictions.
[2] "If I could see farther than the other ones, it is because I was sitting on the shoulders of the giants who preceded me".
[3] This illustrates the power of both the concepts and the mathematics which use them!

domain of cost could therefore be misleading and rejected, although it correctly describes the human look at the things.

2. The second reason is that – as it was said by Karl Popper – a theory can neither be considered as finished: it has always to be checked with the results of nature and just one phenomenon which does not fit with the theory seriously questions its validity: remember the experience carried out by Morley and Michelson, or the advance of Mercury perihelion. One single experience can force people to adopt another theory. But in the current language, theory is considered as the truth and, again, the common word could be misleading in the domain of cost.

3. The third reason is related to semantics: in the ordinary language, the word "theory" has two opposite meanings. First of all it is used, with great respect, to qualify the work of the giants who preceded us. But the second usage is rather dangerous: if you arrive in a meeting with a cost estimate adding that it was prepared with such or such theory, you may be sure that somebody will demolish your estimate, saying it is just a "theoretical" approach ... The word "model" is much more accepted than the word "theory" and we will use it.

# 11 Definition of a General Model: Example

## Summary

This chapter introduces the subject of the general models.

These models are different from the specific models in the sense that a specific model is built for a dedicated product family and that it can only estimate the cost of a product belonging to the same product family. Whereas a general model is designed to estimate the cost of anything, without making reference to any product family, even if they require, from some of them, the knowledge of a "reference product" to which the product to be estimated can be compared.

In order to illustrate the concept of a general model, Value Estimator or VE was selected due, on one hand, on its capacities, on the other hand, on its relative simplicity. VE is dedicated to the cost estimate of mechanical products manufacturing (no development does still exist); its conciseness justifies our choice.

## 11.1 Definition of a General Model

In Volume 2 a specific model is defined from the concept of product family. The whole Volume 2 is devoted to the development of a specific model and we repeat here the ideas on which they are based:

- To build a specific model you need to have a set of reliable data originating from existing products, as much as possible manufactured by the same company.[1] The data include the functional and technical definition of the products and their costs (several costs can be saved, from preliminary studies, development, production, etc.).
- The products are grouped into families as homogeneous as possible.
- A specific model is built inside each product family.
- The model itself consists in extracting, from the set of information belonging to the studied product family, what will be used for forecasting the cost of

---

[1] Each manufacturer – and this subject will be discussed in this Part III – has his own "culture", his own way of organizing and managing his plant. This culture creates small (small because competition, in a free economy, does not allow for large ones) cost changes from one manufacturer to another one, even when they produce the same object.

new products belonging to the same product family. A model consists in two things:

1. A formula (what is called in Volume 2 the "dynamic center" of the distribution of the cost inside the product family) giving a relationship between cost and technical variable(s) "describing" the products. We remind here that building a specific model does not need the description of the products, but only the description of the differences between these products; this is completely different and helps explain why specific models are far easier to build than general models. This formula will be used to compute the "nominal" cost of a new product belonging to the same product family from which it originates.
2. The distribution of the residuals around this dynamic center. This distribution provides ways to quantify the predicting quality of this formula, the most useful in practice being the confidence interval of coefficients of the formula.

One could define a general model by saying that it is the "opposite" of a specific model, as its purpose is to be able to estimate any type of product. Opposite has the following meaning:

1. A general model is built without making reference to any particular product or set of products. This is why it is called "general".
2. The model itself consists in a set of relationships which are linked together.
3. This model "explains" the way cost changes when going from one product to another one.
4. The model requires a much complete definition of the product to be estimated than the one which is needed for a specific model (this is a consequence of point 1 above).

Point 3 deserves some attention because it explains the logic on which are based most of the general models. A general model does not compute a cost in a vacuum: nature is often too complex and diverse to allow for going *directly* from the product description to its cost. The idea which can be followed is then the following one: we cannot go directly from the product description to the cost, but we can try to go from another product (called a "reference product") we know the cost of to the cost we are looking for.[2] The algorithms will certainly be simpler to extrapolate from an existing product than to directly generate a cost. How "far" can the product to be estimated from the reference product? First solutions require that both products belongs to the same family, the definition of a family being now larger than it is in Volume 2; a second one, more achieved, does not require that: this distinction will make us able to distinguish different types of general models: the first solutions are obviously easier to implement than the second one.

But one can go one step further: to the use of a reference product one can substitute the concept of "culture", this word being used in the anthropological meaning of "the way to think and act". The culture is really what distinguishes one company from another one.

A general model also computes a nominal cost once inputs are entered. As we will see in the following chapters, the quality of this estimate cannot be quantified the same way as when dealing with specific models.

---

[2] The process of using a "reference product" as a starting point is sometimes called "calibrating" the model. This word is not a very happy choice (calibrating a device means tuning it to a well-known source; here we do not tune the model...) and we will keep the words of "reference product".

Definition of a General Model: Example

Potentially a general model should be able to cost estimate any kind of product, even if it does not belong to any product family and if it has never been made before (first of a kind), even if it is very much innovative. This is the second meaning given to the adjective "general".

## A First Limitation

This adjective "general" has to be more precisely defined. This implies to distinguish, inside all the "products" made by the industry, different major classes (the reader must recall the fact that the word "products" is used in an extremely large sense):

- *Class 1*: All the mechanical, electrical and electronic products, whatever the functions they fulfill, whatever their size, whatever the environment in which they will have to operate. This class is extremely large: it includes most of the consumer products (artistic products, whatever their destination, do not belong to this class ...), the products for the industry (from the products which will be added to commercial products to machines and plants themselves), the aircrafts both civilian and military, the maritime products, the armaments, the satellites, the products for nuclear installations, etc. This class can be broken down into three "subclasses":
  1. Subclass 1a: Pure mechanical items.
  2. Subclass 1b: Mechanical and electrical items.
  3. Subclass 1c: Mechanical, electrical and electronic items.
- *Class 2*: All the software, whatever the functions they fulfill, whatever their size, whatever their quality level.
- *Class 3*: The buildings, whatever their size or their destination (offices, plants, accommodation, etc.), and whatever their quality level, depending on their environment (seismicity, wind, climate, etc.).
- *Class 4*: Civil engineering "products" such as roads, bridges, railroads, etc.
- *Class 5*: Underground activities (tunnels, galleries, wells, etc.).

Inside each class, product families can be built, according to the definition we give in Volume 2 of the "product families".

By "general" we do not mean that one model alone can estimate the cost of any of these products. A general model is devoted to one class of products only. There are several reasons for that: each class has its own vocabulary and way of doing business; the algorithms which are valid for one class are probably not correct for any another class; people who work within a class are probably not interested in other classes.

A model will then be defined as "general" if, inside a given class of products:

1. it is able to estimate the cost of any product belonging to this class without ideally explicitly making reference to another product of the same product family it belongs to;
2. it is able to estimate the cost of innovative products of the same class.

Nevertheless some models incorporate "sub-models" dedicated to several classes. This does not change our definition of "general models" because there is practically no interaction between these sub-models.

We do not mention in our classes models for the chemical industry, because to our knowledge no such model exists on the market: in other words it does not seem possible today to estimate the cost of producing a new molecule only from its definition.

This industry uses models but not directly dedicated to it: cost estimators work at the process level. The process engineers define what kind of equipment is required and estimating, using the models developed for the five classes, is done at the equipment level.

The conclusion at this stage is that a general model does address a particular class of products. But let us go at a higher level: we will see that the structure of the models (not the models themselves), whatever the class they refer to, are rather similar and this point is very important. This means that it is possible to discuss about models in general, without making reference to such or such model; this will be the way which is followed in this part. Chapter 12 proposes a classification of the general model structures. This should encourage and help people to create new models, better and more powerful than our existing models: we are just starting to develop models for cost estimating and there is plenty of room for investigating and applying new ideas.

## About Innovation

We mention that a general model if able to estimate the cost of a very innovative product. What do we mean by innovation?

Innovation has different meanings for different people. We can distinguish three levels of innovation:

1. *Developing a new material*, which employs many people and published papers on this subject are quite frequent in the literature. They are building our future. A very good example nowadays is the effort dedicated to what is called the "nano-technologies".
2. *Developing new processes* for manufacturing present or future items. The technical literature is also full of papers about improving the way we machine materials, the way we weld them, etc.
3. *Developing new products* from existing materials or process. These products either fulfill existing functions in a completely different way[3] (hopefully in a better and less expensive way) or fulfill functions that were not yet satisfied at all. We will return to this subject in Chapter 13 when we discuss the concept of "advanced products".

The cost of the first two innovations cannot be estimated by models. We must not forget that models are based of past experiences: if no past experience does exist, no foundation is available to build on them. The fact that the duration and the cost of past innovations varied by very large amounts convinced us of the difficulty, or the impossibility, of their modelization.

However, the third innovation can be modelized, if this innovation uses existing materials and/or processes. People who are engaged in such innovations may have a very limited knowledge of these materials or processes, which may be completely new, just resulting from the first two innovations: their knowledge may be low or even 0. This means that these people will have to experiment, to learn, to make tests and maybe mistakes, etc. until they get enough information to use these new materials or processes. This "learning" can be modelized.

To conclude this section: existing models can estimate the cost of innovation if the innovation consists of finding a new arrangement of parts built from existing materials using existing manufacturing processes.

---

[3] A good example is given by the following saying: "Edison did not invent the electrical lamp by trying to improve the candle"!

## 11.2 An Example of a General Model: Value Estimator

In order to make clear the ideas evoked in the previous section, we present here a description of a simple, but a powerful, general model. It will illustrate, before we go on, the concepts of what is, compared to a specific model, a general model.

Value estimator (VE) addresses product in class 1, subclass 1a, dedicated to pure mechanical items (no electrical or electronic components). It was originally based on estimating the number of operating hours. It was later on extended to cost by introducing the procurement cost of the materials and the fasteners, plus the hourly rates, and therefore can easily estimate the products costs.

VE was developed in 1990 by Peter Korda for making quick and reliable estimates of products at a very early stage of products developments. The reason[4] for such a model, now generally accepted, is the following one: "low cost must be designed into a product – many studies by industry, government and research groups concluded that, at the time of completing the concept design of a product, 65–75% of the costs of that product have been fixed; furthermore, at the time when the design is completed, drawings are signed off, 85–95% of the product costs have been determined. One cannot expect significant cost savings by investing heavily into automated manufacturing methods, if the product is not designed by the engineer to reflect a highly mechanized manufacturing process".

The purpose of VE is to give the design engineer a tool allowing him/her, without quitting the drawing table or the computer screen, to make an estimate of what he/she sees in front of him/her. This will allow him/her, when a modification of the design is done, to immediately know the change in the cost.

The organization of VE follows the organization of a general model:

- *Product description*: What is it?
- *Environment description*: In which environment is the product going to operate?
- *Manufacturing description*: How are we going to manufacture it?
- *Quantity to be produced*: How many?
- *Computation and results*.

In order to allow the reader to practically follow VE process (VE is included into EstimLab™), actual screens (only the portions of these screens which are useful for cost estimating) are presented.

### 11.2.1 Product Description

The product description is split into three screens:

1. The first one is dedicated to:
    - The quantity to be produced and the expected production rate.
    - The quality level, here defined by the environment in which the product will have to operate.
    - The product size, quantified here by its finished mass (VE can used metric or imperial units). The mass of the slug must be given.
    - The material(s) it is going to be made of. A mix of five materials can be entered; the materials list includes the composite materials. To each material are attached:
        – a name;

---

[4] Several comments included here are extracted from VE user's manual.

- a "machinability" – machinability is defined (Ref. [52], pp. 1–58) as "the relative ease with which a steel is cut by sharp tools in various operations, such as turning, drilling, milling, broaching, threading, reaming or sawing";
- its procurement cost in $/kg – or any other currency (this input is optional; it has to be entered only if the user wants to add this cost to the manufacturing cost) (Figure 11.1).

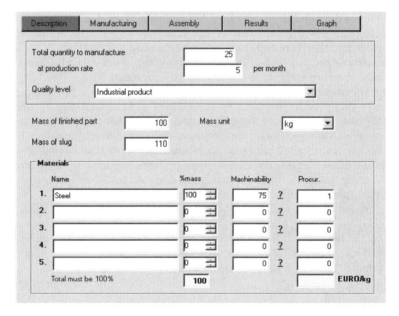

**Figure 11.1**  VE first product description screen.

2. The second screen completes the product definition and give some information about the manufacturing process:
   - Tolerances are entered (they are important cost drivers). To each tolerance (five are available) are attached:
     - the value itself;
     - the percentage of the total surface which is concerned: VE uses here the surface for quantifying the amount of work to be done, which is realistic.
   - If this is the case, the surface finish, in microns if the metric system is used, and the percentage of the surface which is concerned by it.
   - The manufacturing process is shortly described. This manufacturing process includes the way composites are prepared; this is an important element, as composites are more and more used today.
   - If it is required, the user can specify if he/she wants to use a "corrected by constant formula[5] instead of the standard "multiplicative formula". This specification bears the right name of "with adjustment for small masses" (Figure 11.2).
3. The third screen is dedicated to the assembly, if the product is made in more than one piece. We find here:
   - The number of parts per unit and the number of fasteners.

---
[5] The correction by constant formula is defined in Volume 2, Chapter 12.

### Definition of a General Model: Example

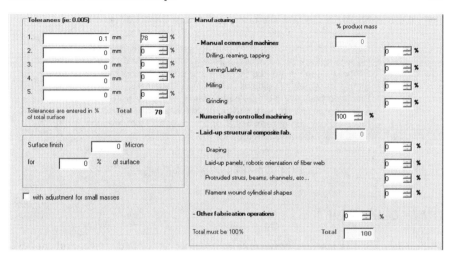

**Figure 11.2** VE second product description screen and manufacturing process.

- The estimated cost and mass of these fasteners (optional, if the user wants to compute the real final mass of the product and include in the cost the procurement cost of the fasteners themselves).
- The assembly process (Figure 11.3).

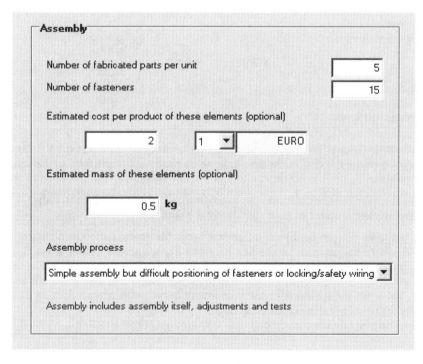

**Figure 11.3** VE third screen.

Several assembly processes are available, as listed in Figure 11.4.

> No assembly required or single part
> Simple assembly, standard fasteners or adhesive bonding
> Simple assembly but difficult positioning of fasteners
> Precision positioning of fab, parts, pinning line drilling
> Precision positioning with moving parts (i.e. servos, motors,...)
> Highest precision assembly (i.e. optics, gauges, instruments)
> Majority of assembly with robotics

**Figure 11.4** VE assembly description choice.

## 11.2.2 Quality Level

The quality level is also an important cost driver as it is in all general models, because such models have to estimate the costs of products made for different people having different requests as far as the quality is concerned.

As this quality level is strongly correlated to the environment, more or less difficult, in which the product will have to operate, which, as its turn, is correlated to the industry type, VE quantified the quality level by the customer requirements; the requirements are grouped into four groups, which are related to the industry type (see Figure 11.1):

- consumer goods,
- industrial goods,
- aircraft,
- space industry.

## 11.2.3 Results

### Hours and Cost Results

Once all the inputs are entered, the number of hours is computed.[6] VE distinguishes, because the cost of the first items to be produced rarely follows the definitive pattern:

- the average number of hours per manufactured unit on the first 10 units,
- the average number of hours for the total quantity.

VE computes independently the number of hours for manufacturing and assembly.

VE allows the user, if he/she has previous experiences, to enter a multiplier which will adjust the model algorithms to the way of doing business of his/her specific company. This procedure – which is recommended if the information is available, but is not compulsory – uses what was called a "reference product". VE then conforms to the logic previously explained.

If an hourly rate is entered costs can be computed (Figure 11.5).

---

[6] Note that VE may use a "correction by constant" formula (see Figure 11.1 where this option appears).

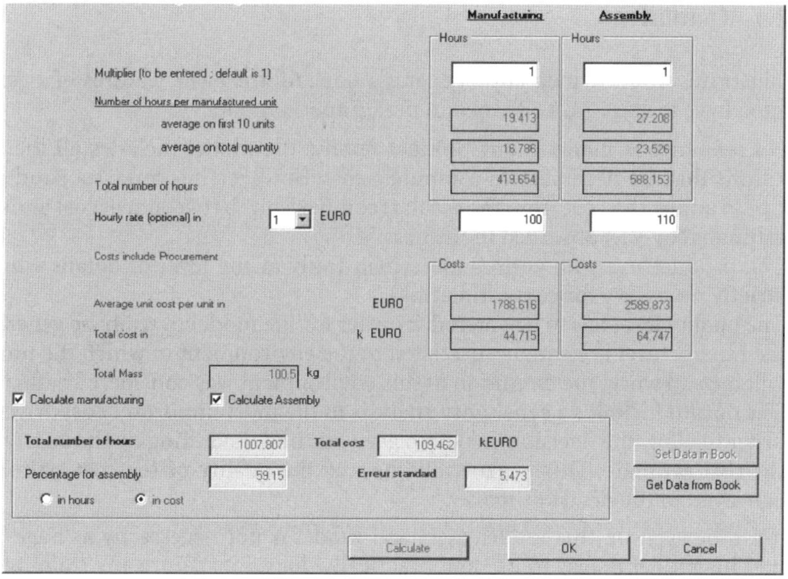

## Definition of a General Model: Example

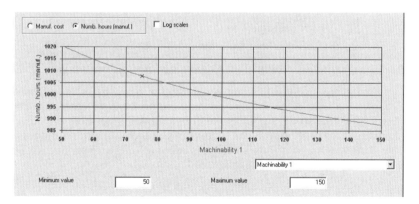

**Figure 11.5** VE results.

### Sensitivity Analysis

As explained later on, general models do not have the possibility to compute confidence intervals for the estimates.

One substitute is to display the sensitivity of the cost to a change of one parameter; this sensitivity immediately informs the decision-maker on the risk he/she makes by accepting the cost estimate. Discussing with the cost estimator on why he/she chose such or such value of the parameter gives some light on what can be expected if this value has to be corrected.

VE delivers this sensitivity analysis on a graph: the example given in Figure 11.6 illustrates the result (given here in hours) when the machinability of the first material goes from 40 to 100.

**Figure 11.6** Sensitivity analysis.

### 11.2.4 Conclusion

VE illustrates what we mean by a general model. All the basic features of a general (limited here to class 1a, as defined in paragraph 1) model:

- No reference is made to any product family: the model includes all the algorithms that are necessary to compute a cost. However, it is always a good practice to adjust the cost from the results (comparisons between real cost and costs estimated by VE) observed in the past.
- The product must be entirely described (only at the level of details which is strictly necessary for cost estimating).
- The quality level has to be entered, in order for the model to really be general. As the quality level is strongly correlated to the environment in which the product will have to work, the definition of this environment was considered sufficient.
- The output includes a sensitivity analysis to the most important cost drivers, in order to allow the decision-maker to get a fair understanding on the accuracy of the cost estimate. This is generally the way the quality of the cost estimate is conveyed to the decision-maker.

As it can be seen on this model, a general model is not necessarily a "huge" program! This model is easy to use and practically does not require any training. The only limitation is that the user must have some understanding (not very detailed) on the type of machine which will probably be used in the manufacturing process.

VE can estimate the cost of any hardware product, large of small, complex or simple, made out of any material.

# 12 Building a General Model

## Summary

We stay here at a purely "academic" level; our purpose is only to describe in general terms the different ways general models can be built, certainly not to make a judgment on existing commercial models. No comment or sentence can be interpreted in terms of "good" or "bad": it is not the purpose of the author to make a judgment on the outputs of any model or even to compare two models.

Any solution created by a model builder may give good answer if it is properly developed and, most importantly, properly used.

In order to explain how general models may be built, a classification of the models is proposed, the Ariane thread between these models being the distance they take to the concept of product families.

The first type of model is a set of many specific models, each one devoted to a "large" product family. In order not to have a too large number of such models, the model builder has to get away from the concept of homogeneity inside each family. The price to pay is to add a subjective parameter which the cost estimator may adjust to his/her own perception of the product.

The second type of model is a generalization of a formula. Chapters 4 and 9 showed that – and this is obvious in the petro-chemical industry – using a "multiplicative formula" is rather successful, but that the exponent depends on the product family and can therefore being considered as a characteristic of the family. Starting from this characteristic may allow to build a general model.

The third type of model abandons completely the concept of product family. Of course what is to be estimated has to be described, but, when this is done, general models of type three look to "who is going to manufacture that" which directly leads to the concept of "culture".

The most important question to be answered when considering building a general model is, once you have decided of course to what type of industry it will be proposed, is: what are the problems the model will help solve? This question supposes a preliminary one: for which phase is the model dedicated to? Is it a model made for the decision-making phase, or for the design engineering phase, or for the manufacturing phase, etc. or for all phases?

The second question to answer is: Which track are you going to follow? This requires to distinguish three types of models. The Ariane thread to distinguish these types is the distance they take from the concept of product family.

We assume in this chapter that all the data we need were normalized according to the procedures described in Part II.

## 12.1 The Three Types of General Models

In order not to be too abstract, examples used in this section will mainly refer to the class 1 of products. Some comments may be sometimes added for the other classes.

Specific models are rather classical: they are part of our culture, as are Kepler's laws, electrical or other physical laws. After all Boyle's law is rather similar to what is called a specific model and built according to the procedures discussed in Volume 2.

Now how can a general model be built?

Three major roads can be followed (the interest of these roads is not our concern here, as we are only interested in the logic of the models and we consider that different users, having different needs, can prefer one or the other):

1. One can first start, inside each major industrial class, by grouping its products into groups, each group including products as similar as possible. The word similar is used here intentionally instead of homogeneous: the purpose is not anymore, as we do for building a specific model, to get homogeneous – and sometimes small – families but to get large groups of only similar products. The problem here is to find a satisfactory compromise between the number of groups and the heterogeneity of each one. One finds here, inside each group, a process rather similar – with some differences however – to the specific modelization, as the user has to "attach" each product he/she wants to estimate – or part of it – to a group, a group being larger than a family. This first class of models is then just an extension of the concept of specific models, the challenge being to take into account the heterogeneity of the products inside each group.
2. The second type of general models is a generalization of a special type of specific model: we saw in Parts 1 and 2 that one efficient solution for building a specific model is to use the "multiplicative" formula:

$$\text{cost} = A \times \text{size}^s$$

and we saw in Chapter 9 that the exponent $s$ was different for each product family. In other words each family is characterized by a special $s$ value, plus of course a value of $A$. If it is possible, once $s$ is known, to establish a relationship allowing to compute the value of $A$ as a function of $s$, one could have a general model based on the following relationship:

$$\text{cost} = A(s) \times \text{size}^s$$

where the cost of any product could be found as soon as its value of $s$ is known. How is this value determined? The answer was given in the preceding chapter: we need a "reference product", belonging to the same family as the product to be estimated (here the homogeneity of the comparison is important) of which size and cost are known; the general relationship allows then to find the value of $s$ and, keeping constant this value – this is the basic hypothesis on which is based this model, hypothesis built on the success of specific models – the cost of the new product is easily computed.

Another way of expressing this logic is to say: the second class of general models modelizes how the cost changes when the size changes, because inside the product family defined by the reference product, only the size is changed.

3. The third type of models is more fundamental: it tries to answer the question: Why is basically a product difficult – and therefore costly – to manufacture? What are the basic reasons of the cost? Is it possible, as Value Estimator (VE) introduced it in the previous chapter, to correlate the product description directly to cost? Obviously this path is completely different from the first ones: these first ones express facts, type three trying to go "below the surface of the facts" in order to understand the cost behavior, whatever the products. Now the concept of product family is completely abandoned, which implies that the model builder has now to describe the product and, from this description, compute a cost. No product family, no reference product are needed anymore, the price to pay being to be able to get a "good" (meaning reliable, not necessarily detailed in order to be able to use the model during the decision phase) description of the product.

We will use the terms "type 1 model" for the first category of model and "type 2 model" for the second one, and "type 3 model" for the last one. Once again there is absolutely no judgment about the efficiency of these types, our purpose, purely academic, being only to suggest to models builders different paths.

In the following sections, we will give some information about the first two types, reserving – due to the novelty of its concepts – the following chapter to the third type.

Before we do it, some general concepts used in all these models types will be shortly introduced. In this chapter the accent is mainly put on hardware cost estimating. The reader will find comments on the other "products" in Chapter 14.

## 12.2 The Price to Pay for Generality

All types of general models aim of course at generality. This implies some constraints, the two important ones being the definition of the size and the quality level.

### 12.2.1 About the Size

We saw about the specific models that the size (meaning: is it a small or large product?) is certainly the most important variable to consider when building a model and that it could be defined according to different ways, technical (mass, volume, length, etc.) or functional (power, load, etc.). It is not possible here anymore.[1]

Suppose the model builder wants to use the "power" as a way to describe the size of product. The fact is he/she does not know how the model user will interpret it: he/she may decide the input power, or the output power, or the electrical power, or the heat power, or the dissipated power, etc.

In order to make a general model, some general size descriptor must be used. It must be recognized that the most general size descriptor for the industry of class 1 is the mass: it is unambiguous, even if it has some limitations. For instance a general model which uses the mass to describe the size cannot be used to estimate the cost of a plant, as nobody knows the mass of a plant, or of a huge heat exchanger (as those used in the nuclear plants)! A second drawback is that the mass is generally a good descriptor of the size of an activity, but not always: polishing the surface of a mirror or painting the surface of an object is obviously not related to its mass. Another example is the surface finish, obviously related to the surface, as does it VE.

---

[1] We will see however in the next chapter that type 3 models can do it.

Nevertheless a model built for the electro-mechanical industry generally uses the mass as the size descriptor, because it is very convenient for most of the products. One thing the user of such a model should not forget: the mass as it used in the model is NOT a descriptor of the product size! It is used as a substitute of the effort to be carried out. If we return to the definition of the cost ("the economic value of the effort to be carried out") it is clear that the cost is related to the effort, not to the product mass. Quite often effort and mass are strongly correlated, but it is not always the case, as illustrated in the examples given in the previous paragraph. Therefore the question that the model user should ask when using the mass is: "can I consider that the effort to be carried out is correlated to the product mass?".

A similar comment can be made for the software industry: the deliverable source instructions (DSI) are often used (they have to be carefully defined) to quantify the software size. It is obvious that the effort to be carried out is strongly correlated to these DSI (not taking into account the code which is automatically generated by the compiler).

A good counter example is given by the micro-electronics industry: nobody would believe that preparing a microchip will be related to its mass. For a given technology, the cost is probably much more related to the surface (something the design engineer knows very quickly by the way) rather than to its mass. The same comment is also true for the nano-technologies, where the size descriptor has still to be found (but the cost data are still too limited to get a definite answer, plus the fact that, at the emergence of a new technology, cost data are very erratic).

We will return to this question in the next following chapters.

In the building industry, the floor surface is the obvious measure of most building sizes, although for large rooms (such as auditorium) the volume might be more adequate.

For a tunnel the length and the section are the natural measures of the size. The volume of rock to be excavated might also be used, but it does not cost the same if the volume has to be excavated on 10 km or 100 m.

### 12.2.2 The Quality Level

Beyond the size – and other characteristics which may quantify the difficulty of manufacturing the product (such as the type of material to be used) – there is always a question to answer: What is the level of quality that the customer requires? The definition of quality is here the standard one: Does the product correctly and for a time to be defined fulfill the functions the customer expects when buying the product?

This quality is defined differently in the different industry classes:

- For *the electro-mechanical industry*, quality is synonymous to reliability in a given environment (the definition of this environment is important: the same product will live longer in a "peaceful" environment than in a "vibrating" environment).
- For *the software industry*, quality is synonymous to the lack of "bugs".
- For *the building industry*, quality generally means stability in a given environment (quantified by the level of seismicity, of wind, of tornado, of rain or sun).
- For *the underground industry*, quality is highly dependent on the nature of the rock to be excavated (fractured, water, etc.). A tunnel built for vehicles is supposed to last a large number of years. This is the major difference between such a tunnel and a mining tunnel for which the lifetime is much shorter. This quality level has a price.

Quality level may be extremely expensive: this is one of the reason which explains why, in the electro-mechanical industry, "space" products are much more expensive than ground product, the other ones being that space products are much more machined in order to save mass, that a clean room must often be used, etc. (the cost ratio is higher than 10). For an aircraft the amount of time spent on software testing is about seven times the amount of time needed for coding it.

In order to be general, a general model must quantify this quality level. A quick analysis convinces model builders that the quality level is very strongly correlated to the severity of the environment: the more difficult is the environment the more quality level is required: the quality level explains the amount of time devoted to controls, testing and traceability. Associating quality level and severity of the environment is however a bit short; we will see another approach in the next chapter.

For software, a similar approach is often taken. This does not seem to be realistic; it should be related to the expected amount of defaults associated with the level of their consequences. But there is a strong correlation between the expected quality level and the industry type: a better quality is required for airplanes than for cars for instance.

For building, the same reasoning is made. This industry is so large that rules have been decided and a quick look at these rules reveals the correlation between the cost and the environment.

### 12.2.3 The Time

The question of the time interests mainly the industry of class 1: a product which costed $100 twenty years ago, may cost less than $50 today, the economic conditions being supposed to be normalized. This phenomenon is very important in this industry and must be considered.

The logic is not difficult to understand, as the cost decreases rather rapidly when it is offered to the market (there are plenty of evidences about it) until the rate of decrease stabilizes to a value between 2% and 3% a year.

## 12.3 Type 1 Models

Type 1 models are still very close to the concept of product families.

Type 1 models can be considered as an extension of specific models. By definition a specific model deals with a product family. As the number of product families would be extremely large if the constraint of homogeneity – on which we insist so much in Volume 2 – was enforced, it has to be relinquished. Therefore the idea of product family is extended to function families, such a family gathering products fulfilling similar functions, even if the product design is different.

The first work of the model builder is to decide about the number of function families he/she will use. This number is a compromise between a large number of families with a small heterogeneity, and a small number with a large heterogeneity.

Once the questions of size, quality level and time are potentially solved, the model builder quickly realizes that the data belonging to one family are too much scattered to be used directly. Then this person adds one (or several) qualitative variable(s) to this family,[2] defined on a limited number of levels and tries to split his/her family into sub-families: this qualitative variable may bear the name of "difficulty" and an average of five levels is sometimes considered as sufficient to obtain acceptable results and to be easy to use. These five levels may be called: "very difficult", "difficult", up to "simple". It is clear that this variable is in fact a subjective variable.

Chapter 11 of Volume 2 explains that using a qualitative parameter only translates the formula in the case of an additive formula – multiplies it in the case of a multiplicative formula – and therefore the process should not be too complex. However, it has to be checked, as we do it for the qualitative parameters: it is quite possible that the slope has also to be adjusted, which means that five formulae may have to be generated.

This procedure is theoretically good; it has however an important drawback, related to the subjective parameter, which will be discussed in Chapter 17.

Using such a model is not difficult. The user selects the family he/she wants to work with and for this family decides about the level of difficulty. As previously explained, the model builder must help the cost estimator by giving examples of what he/she means by "very difficult" ... The major problem here is that the model builder does not know who is going to use the model and for which type of industry. Finding examples that the cost estimator will understand is not too difficult if the model builder and the cost estimator belong to the same company, but may be difficult in other situations ...

For this reason it is always a good precaution that the cost estimator, using his/her own "reference product", builds his/her own scale of difficulty. The results will certainly be more credible!

## 12.4  Type 2 Models

Type 2 models take some distance with the concept of product families. This concept is still useful, but it becomes less important here.

These general models are built on a formula such as:

$$\text{cost} = A(s) \times \text{size}^s$$

where $s$ is the exponent we met in Parts 1 and 2. This formula means that the specific cost of product (their cost per unit of mass) follows a curve as illustrated in Figure 12.1: for a given exponent $s$, $A(s)$ remains constant and the shape of the cost depends only on size$^s$. Figure 12.1 is just a copy of what was said in Chapter 9 about the Chilton's law:

$$\text{specific cost} = A(s) \times \text{size}^{s-1}$$

What could be the function $A(s)$? We saw in Chapter 9 that, according to the values found in the petro-chemical industry, $s$ remains less than one and that it becomes closer to one as soon as the equipment becomes more difficult to manufacture: $s$ could therefore be seen as an indicator of the difficulty to manufacture. In such a case we can expect that $A(s)$ is a function which increases with $s$.

---

[2] A good example is given by COCOMO developed by Barry Boehm [3].

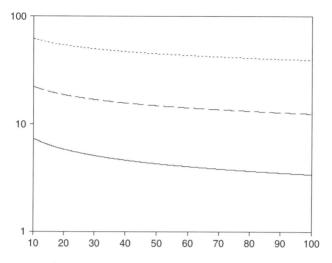

**Figure 12.1** Specific cost for different product families.

## 12.4.1 Type 2 Models: Distinguishing Between "Laws" and "Initial Conditions"

We can introduce here a few comments about model building in general. In science two types of models do exist. Let us take the example in astronomy:

- Kepler, starting with a lot of observations about the positions of the planets, was able to establish the three laws which bear his name. These laws were remarkable, they are still used, but were limited to the motion of the planets and did not "explain" the basic reason for these particular laws. His work described facts. We would say nowadays that they were valid for a "product family".
- Newton established a more fundamental law from which the Kepler's laws could be derived. Newton's law was more fundamental and could be applied to any time of motion in a gravity potential.

Why are Newton's law more powerful than Kepler's laws?

As the same logic is followed by all the sciences (going from formulae describing the facts to "conceptual" models), there must be a fundamental reason to do it!

The basic reason which has been found by the human mind, centuries ago, is to say: let us distinguish between the general laws and the initial conditions.

### About the Laws

The general laws do not refer to any particular situation and this explains why they are so difficult to find out: they require a large amount of abstraction and the invention of concepts. The role of these concepts is to "explain" the general behavior of the observed phenomena: the word "explain" is voluntarily put inside quotation marks because, as it is well known, it is inappropriate; the role of the concepts is only to be able to predict the general behavior of a system from abstract considerations. Examples of these concepts are numerous: forces for Newton, fields for Maxwell, quanta for Planck, entropy for Boltzman, etc.

The problem is then to link these concepts with observable quantities.

Let us illustrate that by the law of gravitation. Forces were considered before Newton but the merit of Newton was to link them with quantities which can be observed, here the mass $m$ and the acceleration $\gamma$ of a body, through the most powerful and simple relationship ever found:

$$\vec{F} = m\vec{\gamma}$$

The beauty of this relationship comes from two points:

1. this simple relationship allows to predict the motion of anything when the forces are known;
2. and, as Newton said, forces are easier to manipulate than motions.

It is important to mention here that the generation of these laws has nothing to do with the process of "extracting" from data some relationship able to be used for forecasting. This is why we often call these laws as "conceptual" models: they are built in the model builder's mind (think about the mental experiments of Albert Einstein) who eventually considers that nature should obey such laws, even if no experiments has been done yet for testing it. Of course the model builder is deeply influenced by known past experiences, but he/she builds his/her model on ideas which could explain the results already observed.

## About the Initial Conditions

Laws by themselves do not allow to make any forecast. The forecast can be done if the initial conditions are known: it is the couple "laws + initial conditions" which allows to make predictions about a phenomenon.

This separation between general laws and initial conditions is one of the most remarkable idea developed by the human mind: it allows to replace an infinite diversity (what we observe in nature) by something simple. It does not mean that no effort is still necessary, because we have now to compute or estimate the value of the concepts. For Newton, it was the "law" of universal attraction between bodies.

This discovery is so important that nobody can consider building a general model in any science without using it. How does it apply to models for cost estimating? This explained in Figure 12.2.

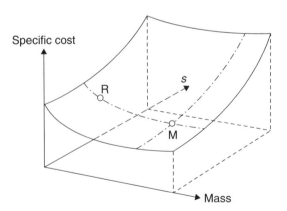

**Figure 12.2** The idea of a general model based on the specific cost.

The surface represented in Figure 12.2 is the general model: each product, defined by its specific cost and its mass, can be located on this surface. Now how can we predict the cost of a new product? We know its mass and therefore the cost will be on the mixed line corresponding to this mass; in order to locate it on this curve we use a reference product R (our initial conditions) for which the mass and $s$ are known: it defines on the surface a line with constant $s$. This line will intersect the previous one at a point M which gives the specific cost of the new product.

## 12.4.2   A Few Difficulties When Using a Type 2 Model

As said earlier, a type 2 model assumes that the mass is a correct substitute to the size of the effort to be carried out. We will see in the next chapter that one basic problem will have to be solved when products built from different materials have to be compared. But we presently discuss another problem with which any cost estimator has to cope every day.

### About the Products

A type 2 model is based on the hypothesis that the exponent $s$ remains the same inside a product family: therefore the exponent found for the "reference product" (our initial conditions) is supposed to remain constant and is used for estimating the cost of the product to be estimated. As largely mentioned in Volume 2, this is true only if the reference product and the product to be estimated belong to the same *homogeneous* family. It is not always the case: the fact that both products bear the same name is not a guarantee that they strictly belong to the same family.

In Volume 2 the problem is exactly the same and the solution is found by adding "secondary" (the "primary" parameter being the size) parameters.

This solution is the only one. Consequently in a type 2 model the cost estimator should be able to adjust either the exponent or the factor $A$ in order to take into account any difference between products. Models builders should investigate this possibility.[3]

### About the Size

As previously mentioned, general models of type 2 use the mass (for hardware) or the number of instructions (for software).

The general formula does not say anything about its domain of validity. The user must check it; is quite possible that, when the size changes, the exponent has also to change, due to the use of a different technology.

### About the Orthogonality Between Mass and Exponent

This is a problem related to the previous one: the general model assumes that both variables are "orthogonal". This means that the exponent and the function $A(s)$ are supposed to be independent: changing the mass does not change the exponent.

This obviously has to be checked.

---

[3] Chauvel [18] gives numerous examples of this technique.

### 12.4.3 Conclusion

Type 2 models may be an efficient way to build models. But the reader must check that the assumptions on which it is based are fulfilled.

One can also add that the formula called "correction by constant" should probably be used if products with a small mass have to be cost estimated. This type of formula is explained in Volume 2.

## 12.5 Type 3 Models

Building a type 3 model offers interesting comments on the process of building models, whatever the domain, including science.

The logic of the construction of a general model is represented on Figure 12.3. This logic is common to all sciences and is therefore not original. It includes several steps which are briefly commented:

- Everything starts from data. Data are then split into two groups: one set of data will be used for fixing the "constants" of the model, the second set for model validation.
- Now comes the most important part of a general model building, which cannot be reproduced because it depends too much of the intuition of the cost analyst. This person, by experience, by data analysis, maybe by computing preliminary formulae, by theoretical considerations, by just thinking about the data (the thought experiments of Albert Einstein …), etc. builds tentative relationships which describe how cost should behave, in general terms, which means without

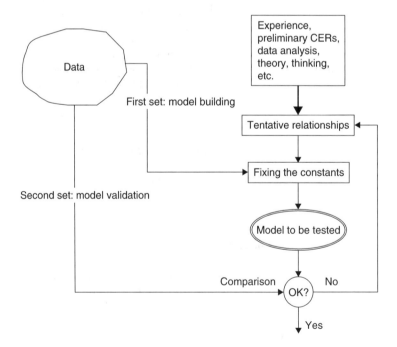

**Figure 12.3** The logic of model building. CER: cost-estimating relationship.

referring to a special or specific situation. They can be called "theoretical relationships" because they explain how, in this person's mind, costs behave the way they do. These relationships always involve creating concepts, or "invisible ideas" from which the cost behavior may be inferred.
- Once these tentative relationships are established, the constants of the model must be quantified. Any model uses constants which we are, presently, unable to quantify by pure reasoning: for Maxwell it was the speed of light, for Newton it was the constant $g$ of gravitation, for Einstein it was the cosmologic constant which is so much discussed nowadays, etc. The only way we can put a value on these constant is, at a given stage of theoretical development, by observation. Observations which are used here come from the first group of data. Once this is done, the "tentative" model is ready to be tested.
- Now comes the most critical, and sometimes disappointing, part of the process: testing the model. The model is tested with the second group of data. If the comparison between the values computed by the tentative model and the observed values is satisfactory, the model becomes ready to be used; practically a lot of other developments have to be done to make it usable, but the "heart" is there. If the comparison is not good, the model builder returns to his/her thought until he/she gets a new idea from which another tentative model can be prepared.

The whole process may take years. Quite obviously it is completely different from building a specific model.

Once this is done, it is possible to develop algorithms for other costs related to the same product family.

## 12.6 Other Points

Up to now only the manufacturing cost was mentioned. Obviously all types of models include other cost-estimating procedures for tooling, management, documentation, development, integration, etc.

From an academic perspective, the manufacturing is one of the most interesting feature (for hardware) and this is the reason why it was used in the presentation of the different models.

# 13 New Concepts in General Modelization for Hardware

## Summary

The electro-mechanical industry was, due to the amount of data collected for generations in this industry, the first one who saw the emergence of general models.

We already presented, in Chapter 11, "Value Estimator" (VE) which was used as an example in order to introduce the concept of general models.

The purpose of this book is not to make a presentation of all the existing models in the hardware industry and certainly not to compare them and the results they give; all of them are sold and any information about them can be obtained from the sellers. Unfortunately models builders are generally very eloquent to describe the beautiful sides of their models, not so much about how they were built, and especially why they were built the way they are, which is precisely what we are interested in here.

Consequently we will concentrate here on some concepts which were recently introduced: they might first interest the community of cost analysts, offer a track for education or for further research (models are far from being completely achieved) and, why not, the realization of other models.

The chapter starts by a presentation of what can be the description of a product for cost-estimating purposes. The reader must not forget that, if no comparison is possible with any other existing product, then the product has to be described in "absolute terms" ("relative terms" means that we compare the product with other products). Product description is therefore a very important subject when building a general model.

Two points are explored: the description of the product **size** on one hand, the description of the **other aspects** of the product on the other hand. The first point mentions the use of the (external) physical characteristics or the (external) functional characteristics. The second point studies the very important concept of the internal technical functions (ITF), from which the product structure, which synthesizes the product description beyond the size, is derived.

Then the quality level is deeply investigated from the requirements, built on the distance to the elastic limit the material(s) have to work.

After discussing the materials, the type of formula to be considered and the type of production are investigated.

Eventually the concept of "organizational structure" is developed as it is the key to be able to cost estimate without referring to a product family or even to a reference product.

## 13.1 Introduction

This chapter develops a few concepts on which a type 3 model can be built. The reader is reminded that the definition of a hardware type 3 model has two components:

- The model must be able to estimate the cost of any type or hardware (mechanical, electrical and/or electronics) whatever its size, the domain of usage, the quantity to be produced and the way it is produced. For instance the model may have to estimate products weighting less than 10 g, produced at a quantity over 100 million (in a highly automated plant; this has already successfully been done), or be used for medium or large product (up to 1000 metric tons), this time for one unit only (this has also already been done).
- The model should not be obliged to use a "reference product", which means that the concept is abandoned. Nevertheless it should be able to use such a product, if one is available even if it is completely different from the product to be estimated and if the cost estimator thinks it can improve the quality or at least the *credibility* of the estimate. As credibility is very important in the cost-estimating domain, anything which can improve it deserves our attention.

A model using the concepts described in this chapter has already been implemented. Our objective is not here to describe this model, but, remaining at an academic point of view, to present these concepts as several of them are highly innovative. They may help the reader to create new models if he/she is interested in such a task.

As we believe that the major objective of a cost model should be for helping decision during the decision phase, we concentrate here on this first focus. Nevertheless the model should also be able to estimate when more information is available, during most of the validation phase.

The challenge here is that the amount of information about the product is generally limited.

We just said in the introduction to this chapter that a general model does not need the knowledge of a reference product. As a matter of fact, we have different situations which can be summarized by two extremes; these situations illustrate ideas which are strongly correlated: the level of information which is required on one hand, the existence of a reference product on another hand:

- Suppose there exists a reference product belonging to the same product family. After all this is, for the cost estimator, a comfortable situation: he/she just has to extrapolate from this reference product. This is the way type 2 models are built.
- But we may also have a reference product which does not belong to the same product family and we would like to use its known cost for increasing the credibility of our estimate. The easiness of the situation depends now on the distance between this reference product and the product to be estimated; if this distance is a bit large both products have to be described in order to compare their costs. The description can be here a "relative" description: starting for the reference product, we just have to describe how the product to be estimated differs from it. It may have also to be described in "absolute" terms if the distance between products is too large.
- If there is no reference product, the cost estimator has to palliate this lack by another information, which is here a description – in absolute terms – of the product to be estimated, plus another information related to the way companies

make business in the domain, what we call the industrial culture. This is rather new as it was not strictly needed for types 1 and 2 models.

A general model must cope with all situations. We will mainly discuss the third one, but the first two must not be forgotten.

To cope with these situations, a general model needs a description of the product to be estimated: saying that the level of information is limited does not mean that the model can work with no information … The information which is required by the model depends on the possibility or not to compare it to another product, belonging to the same product family (the concept of product family can be still used here!). As you may expect, this level of information must be more "detailed" if no comparison can be made. We will start with this hypothesis.

Information includes in such a case (no comparison is possible) two major cost drivers, plus more "conventional" ones such as the material(s) to be used. These cost drivers are called here the "product size" and the "product structure".

Let us add just a word about the word "function" which will be used several times in this chapter. This word will refers to two different concepts which must be clearly distinguished:

- Either to the external technical functions (ETF) the product is supposed to fulfill for its user. We will find for instance these ETF when discussing the size of the product, because the size is often, especially at the very beginning of a project, quantified in terms of these ETF – for instance so many kilowatts, so many seats, so many cubic meters per second, etc. Another word which could be used instead of this ETF is the external functional capability of the product (external means this is the point of view of the product user).
- Or to the internal technical functions (ITF) which are generally unknown to the user, but are the specific solution (let us hope it is the most efficient one!) selected by the design engineers. Another word, frequently used although its definition is not so clear, is "technology".

We immediately see from this introduction that we must, as cost analysts, have two different views of the product:

1. An *external view*, as the product user sees it.
2. An *internal view*, as the product designer sees it.

These two views are not opposite; they are complementary and needed for cost estimating.

## 13.2　Describing the Product

### 13.2.1　The Product Size

The product size, as it is explained in Volume 2 about the specific models, is a very important cost driver. Some attention should therefore be devoted to its quantification.

First of all the product size has to be defined from an external point of view. In order to define this size, we must consider we are looking to the product, as its user would do.

There are basically two ways to quantify a product size:

1. Using some *physical descriptor*, the most common ones being the mass or the volume or even the surface (VE used both mass and the surface). This is generally the cost estimator's point of view for cultural reasons (technical descriptors are easy to compare from one product to another one).
2. Using a *functional approach*, this is an answer to the questions: What is the product doing and at what level? This is the customer's point of view.

Ideally, it would be interesting sometimes to go from the second to the first. The customer, mainly for mobile products, is interested in both the functional characteristics and the physical ones.

In most situations, we consider that the model's user should be able to decide on the descriptor he/she prefers for the particular problem to be solved. What the model builder can do, as far as the functional characteristics are concerned, is the following:

- Either to propose built-in characteristics, each time he/she is able to include them. For instance for customers working with rocket engines, the thrust can be an interesting one, especially in the early stages of a project. Therefore the model builder is encouraged to add to the model several, as many as possible, functional characteristics. The real problem this person is faced to is to have an unambiguous definition of the characteristic.
- Or to let the user use his/her own characteristic: it would be impossible to include in a model all the functional characteristics a customer would be interested in. Therefore a procedure has to be available inside the model for letting the user use his/her own set of characteristics; there is no reason to limit this procedure to only one characteristic, as the model user could be interested in working with several at the same time. To return to the example of the rocket engines, the user may want to work simultaneously with the thrust and the specific impulse. This is quite possible. The price to pay, because there is always one, is that the user should possess a few examples of such characteristics related to cost: the model cannot "guess" how a peculiar characteristic can change the cost but it can build the relationship between them if a few examples are available. This means that as far the product size is concerned we are back to the concept of product family.

Nevertheless a general model can help him/her in several ways:

- he/she needs a very limited set of data, because all the necessary equations are built in the model (whereas building a specific model requires a lot of data).
- procedures remain available to introduce changes (such as materials or quality or any other ones, plus the estimate of the development, maintenance, ... costs for which building a specific model would require new data).

## Using a Physical Characteristic

Three "global" physical characteristics are in competition: the mass, the volume and the surface.

The mass is generally preferred for products which have to be "moved" or displaced, because the cost of this displacement may be important for the project (such as in space activities) or to the user (the mass for instance drives the gas consumption of a car). The mass is also often preferred for small products because it is not so difficult to estimate it, even in the early stages of a project.

However, for large, fix products the only information which is available is the volume: for instance for describing the size of a refrigeration tower for a nuclear plant, engineers have, very early, a good idea of its volume and certainly not of its mass (during the decision phase). Using the volume for describing the size of small products is however generally to be avoided, because manufacturers quite often normalize, for obvious manufacturing reasons, the volume of their products: the same volume can therefore contain more or less equipment. This happens frequently for electronic boxes.

The mass has a competitive advantage on the volume because there is only one way to quantify the mass, whereas there are several definitions of the volume. Nevertheless two people discussing about the volume of a refrigeration tower for a nuclear plant would certainly agree on the definition of the volume.

Surface is rarely used by cost estimators, except for very special equipments, such as optics. Nevertheless it should be considered, as it is in VE (see Chapter 11): after all when machining an object, it is the surface which is machined, not the volume or the mass. One objection which could be made in the past is that the surface may be difficult to get; it is certainly true at the very beginning of a project – and there the mass or the volume have an advantage – but once the CAD (computer-aided design) works on the product, the surface is easily given as an output.

*A General Discussion About the Use of the Mass*
As mentioned in the previous chapter, the mass is used for describing the size of the effort to be carried out – or, if you prefer, is used as a substitute of this size. This does not create a particular problem as long as the same material is used for all the products you want to compare. But it is not always the case, as some products may be manufactured from different materials: suppose you have a product you would like to use as a reference product and that this product is made out of steel, and that your company, for obvious reasons, want to design a new product exactly the same,[1] but this time made in aluminum.

First of all you will have to take into account that aluminum is easier to machine, and that its procurement cost is different. This question will be answered to below. Now the mass is used for quantifying the product size; as the new mass is smaller than it could have been if the product were made out of steel, the model will "consider" that it is a small product and cost it accordingly.

This is certainly not right: suppose the new product is exactly the same as a previous product made out of steel (this is not generally true of course due to the different characteristics of the materials, but this hypothesis is made just to explain the situation); we have already taken into account that the procurement cost and the difficulty to machine the material are different. If we use now the mass as it is, we certainly "lie" to the model.

One solution is to convert the mass of the new product **as if** it were made out of steel, for exactly the same size. Now, the comparison between the reference product and the new product is realistic.

Obviously, the problem would not arise if the volume is used for quantifying the size.

---

[1] This is just an example: if there is a change in the material, both products may not have the same size in order to fulfill the same functional characteristics. But, at this stage and for explaining the concept, let us assume both products are identical.

## Using Technical Characteristic(s)

Using technical characteristics (see my article in ISPA, 2003) (or ETF) such as the power, or the load, or the frequency, or the range, or anything else ... would be a good idea during the decision phase of the project. By ETF we mean the characteristic(s) the user of the product is concerned about.

It is a good idea for a lot of reasons:

- It is the first characteristic(s) to be known in most projects and it is always a good idea to estimate from this first characteristic: being forced to compute a mass is time consuming, needs resources (which will be wasted when the first modification occurs and modifications are frequent during the decision phase), requires making hypotheses and is therefore necessarily inaccurate.
- The person who receives the estimate must understand the relationship between the cost and the technical characteristic(s) as well as the sensitivity of the cost to a small change in the characteristic(s), at least to be able to get an idea by him/herself about the reliability of the estimate: if a small change of a characteristic implies a large change in the cost, he/she should worry about the accuracy of the cost estimate.
- This person has always to make trade-off analysis: What is going to happen, as far the cost is concerned, if this characteristic is alleviated? What about reducing this specification and increasing something else?, etc. If a mass has to be computed each time a new concept is considered, the process will take a lot of time...
- He/she has also to negotiate with the customer; this customer does not care, except in special circumstances, about the mass: he/she is mainly, if not only, concerned by the technical specifications of the product.

The question here is to understand the need of the person who receives the estimate. This person is of course interested by the cost, but certainly also, about the influence of a change in the characteristic(s) on the cost; quite often in the decision phase, the second is more important than the first.

## From the Technical Characteristics to the Mass

At the very beginning of a project, generally only technical specifications are known. As explained in the previous section, it is then possible to cost estimate directly from these specifications.

It may also happen that the mass is an important characteristic of a product, especially if the product has to be "moved" such as a satellite for which, obviously the mass must be known as soon as possible (the launch into orbit is always a very important cost portion in a space project). But the mass can also be a primary focus for designing a project: if a new vehicle is considered for instance, the mass it will have if it is designed the usual way should be known as soon as possible: if it is too high, it may be necessary to completely change the design, to force the use of different materials, etc.

Consequently, when a project is still at the decision phase, having a reliable information about the mass can be important. All the procedures developed in Volume 2 could be used to create a formula linking the functional characteristics and the mass.

## 13.2.2 Product Description Beyond the Size: The ITF

The product size is just an element of the product description. As previously mentioned, in a really general model, the product should be "fully" (only for what is necessary for cost estimating purposes) described. This section presents what can be done to get a sufficient description. This description should "speak the language" of the design engineer and use only quantitative parameters; subjective parameters should be completely avoided for obvious reasons. We do not ask the design engineer if the product is "complex" or not but only to describe it as he/she sees it, using his/her own language.

What the design engineer knows is the set of internal functions which he/she decided to use in order to build a product able to fulfill the external function(s) the product is manufactured for.

This section deals with these ITFs. Basically the concept works in two times:

1. The design engineer is asked to quantify the importance of the basic ITFs in the product. This means that a set of basic ITFs must be proposed to him/her.
2. From these quantifications, a synthesis is made. This synthesis quantifies a variable called the "product structure".

The purpose of this section is to indicate that the "product structure" is not an add-on cosmetic subject but is really the basis on which the model can be built. For the cost estimator, the interest is to be able to describe, for cost-estimating purposes, any product even it was never made before, even if no drawing is still available, as soon as the customer or the designer is able to list the ITFs he/she intends to include in the product. Of course if drawings are available, the information can easily be used to quantify these ITFs. But basically the idea is to be able to work before any drawing is available.

*Note*: We discuss here the problem of the cost analyst who cannot compare the product to be estimated to another product, as it was mentioned in the introduction of this chapter. If such a comparison is possible, there is no need to describe the product.

### The Set of Elementary ITFs

#### The Theoretical Approach

The idea of the product structure comes from ideas developed by Lucien Géminard,[2] ideas which are briefly presented below.

Frank Freiman added to Lucien Géminard's concepts that any product is made to "manipulate" energy,[3] under any form. For instance a table can sustain a weight of 100 kg at an altitude of 1 m: the table is then able to manipulate about 1 kJ of potential energy (100 kg $\times$ 1 m $\times$ 9.81 m/s$^2$, the last factor being the acceleration of gravity): this is a very basic form of energy, known for millions of years.

On the other side a transistor junction manipulates a very tiny electrical energy. This is a new form of energy.

Comparing Products Fulfilling the Same Function   The idea goes on with the definition of "advanced product": a product is considered as more advanced than another one if

---
[2] Lucien Géminard, Logique et Technologie, Dunod, Paris, 1970.
[3] The dimension of energy is ML$^2$T$^{-2}$. It is expressed in joules.

it is able to manipulate more energy by unit of mass. A transistor manipulates much more energy by unit of mass than a table: it is a much more advanced product.

How do we make a product able to manipulate more energy by unit of mass? There are two basic ways:

1. The first one is to use a more efficient material: a table made out of aluminum weights, for the same capability, less than a table made out of steel. A table made out of carbon fiber will be a more efficient product, in terms of energy manipulation, than a table made out of aluminum.
2. The second one is to work the material more intimately: a transistor is able to manipulate more energy per unit of mass because a transistor is built by working the material at the micron level, or even less, whereas a table uses material work at about the millimeter level.

In both the cases the cause is always the same: we have to add more work to the product. A more efficient material is more costly than another one, because it demands more work to be manufactured (see table in Figure 13.8 giving the amount of work which is needed to produce different materials); a more efficient product requiring to work the material more intimately includes more work per unit of mass than another one.

In both the situations we have to add work to the material, and this is the reason why the product is more costly to produce. In thermodynamics terms, one could say that in order to make a more advanced product we have to decrease its entropy.[4] It is well known that the change of the entropy of a system is given by:

$$dS = d_e S + d_i S$$

where $d_e S$ is related to the exchanges between the system being studied and the external world and $d_i S$, the change in the internal entropy (this change is mainly related to the uniformization of the temperature inside the system). The entropy of a machine part is lower than the entropy of the raw material it comes from: the amount of "order" is higher: this decrease of entropy comes from the amount of work it receives. A transistor has an entropy, per unit of mass, much lower than a mechanical part and it is the basic reason why it is able to manipulate more energy per unit of mass.

The only way we can decrease entropy is by adding work (the work has the dimension of an energy), or, and this is equivalent, by adding energy. In any case, increasing the amount of energy is the focal point of advancing the state of the art.

**Energy is the common denominator of everything** ... and it is therefore normal to find it in the cost-estimating process.

Comparing Product Fulfilling Different Functions   One can add another point about products. There are two kinds of products:

1. Some products are made just to sustain energy: we ask them to keep the energy until we need it. This the case for a table, for a seat, for a container, for the most part of a kinetic wheel, for cables, for capacitors, for a battery, etc. The history of technologies shows this was the first use of materials by men.
2. Other products are made to change the energy from one form to another form. For instance a crane transforms electrical energy into potential energy, a gear box transforms one type of mechanical energy into another kind, a transistor transforms one kind of electrical energy (direct current) into another kind, etc.

---

[4] The dimension of entropy is in Joule per degree, or J/K.

An important difference between both sets of products is that no change of energy form can be accomplished without a loss; this loss appears as heat and this consequently always requires, even it is too small and achieved by natural convection, some special device in order to get rid of it. Examples are numerous: electrical transformers (which change electrical energy from one form to another one) require some heat transfer, as do mechanical engines (transforming chemical energy, or electrical energy), power transistors (the device may be a small radiator), etc.

An Historical Perspective   The interested reader may establish a historical perspective about the ITF developed by men, from the very basic ones to the more complex ones. The history of humanity can be read on the large book of applications of energy.

*The Practical Approach*
The practical approach was developed by Lucien Géminard in his book. Lucien Géminard wrote for the professors who had to speak about technology, which he defines as "the study of techniques, tools, equipments, materials which are used for a definite action". Studying an object, we would say a "product", he looks at the various technical functions which are used inside this object: these are what was called upward the ITF. He then defines a technical function as something establishing a correspondence between two families of phenomena (cinematic, mechanical, electrical, ...) and says that this correspondence is realized by technical operators, which are the specific parts – although he recommends not to assimilate technical operators and parts.

After this introduction, he considers two categories of functions, which he calls "basic technical functions" and "technical functions requiring several basic technical functions".

He distinguishes seven basic technical functions:

1. Support.
2. Positioning.
3. Containment.
4. Linking of two parts.
5. Motion (which he names "guidance").
6. Control of passive resistances (which is accomplished by surface finish and/or lubrication).
7. Tightness.

A hardware model has to deal with any kind (mechanical, or electro-mechanical or electronic) of product. Starting from Lucien Géminard's approach, it is therefore necessary to add several other technical functions. Basically one may consider 12 elementary ITFs, this number being a workable compromise between detailed and global demands. These elementary ITFs can be grouped into four classes as described below; in this list of ITFs the hierarchy is followed (you can interpret this hierarchy as representing the relative difficulty to produce the ITFs):

1. *Mechanical ITF that do not require changing* the form of energy:
   – Mere transmission of efforts.
   – Containment.
2. *Mechanical ITF mainly and generally involving a change* in the form of energy:
   – Heat (as explained upward).

- Positioning. Three degrees of precision (expandable if necessary) are necessary:
    Standard (about 1/10 mm).
    High (about 0.01 mm).
    Very high (about 1 μm).
- Motion with the same degrees of precision (motion requires, beyond the precision, surface finish, flatness or roundness, etc.).

3. *Electrical ITF*:
    - Not involving a change in the form of energy (cables, waveguides, capacitors, boards, …). This ITF is to electrical ITF what "mere transmission of efforts" is to mechanical ITF.
    - Involving a change in the form of energy: winding (changing electrical energy to magnetic energy), antenna, …
4. *Electronic ITF*: Electronic is a special subject. Due to the inherent variety of components – the basic cost of these components depends a lot, and this can easily be forecasted, on the manufacturing precision – electronics can be dealt with as a whole (this is the case if the electronic is conventional and represents only a small part of the product) or we may have to go down to the nature of the components and boards; the logic is still the same: we have to understand how advanced is the product.

## From the ITFs to the Product Structure

The product structure does follow Lucien Géminard's approach when he mentioned "technical functions requiring several technical functions". The product structure is then a synthesis of the various ITF selected by the cost estimator.

### 13.2.3 Quantification

For implementing the idea of ITF plus their association for making the product structure, in order to make from these ideas a practical tool, two things must be done:

- Defining a value for each ITF. This can be done, now by statistical analysis based on facts, respecting the two principles mentioned in Chapter 2 of Volume 2:
    1. The ITF values should follow a "natural" scale: the more intimately the material is worked (or, historically speaking, the more recent is an ITF), the highest the ITF.
    2. The scale should be about linear. Linear in what? As we are dealing with cost, a practical tool should have a scale which, approximately, follows linearly the behavior of the cost: this scale can then become "intuitive" to the cost estimator.
    
    Once these principles are given, the value to give to each ITF must come from the world. This is why the construction of any model requires, as illustrated in Figure 12.3, some data.
- Defining an algorithm for quantifying, from a mix of various ITF, the product structure. This algorithm cannot be linear as a simple example illustrates.

Let us consider two glasses of the same size (let us say the same mass) (Figure 13.1).

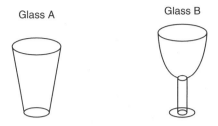

**Figure 13.1** The problem of two glasses.

Glass A is a pure "container", whereas glass B is a mix of "container" + "support". As "support" is a lower ITF than "container", a simple mix of ITFs for glass B would quantify its ITF as lower than for glass A; this would mean – according to the rule selected for the quantification of the ITF – that glass B would be cheaper than glass A. Intuitively this is not the case and your intuition is correct: glass B is more expensive to produce than glass A and, consequently, its product structure must be higher. This, extremely simple, example shows that the algorithm for quantifying the product structure must be non-linear.

*A Geometrical Perspective*

Let us consider a space with 12 dimensions (the number of the ITFs). Any product, of which the description of its ITF is known, can be represented in this space as a point P as illustrated by Figure 13.2.

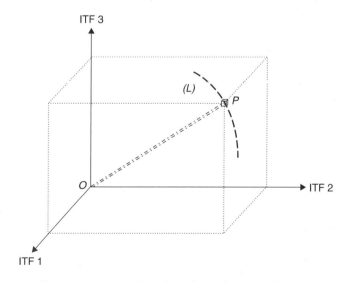

**Figure 13.2** Representation of a product in the space of the ITFs.

In this space the length OP (the double dotted line on the figure) can be interpreted as the size of the product. With this idea in our mind, we can go one step further and try to answer the following question: How a product family (which is the subject of the whole Volume 2) can be represented in this space? Let us assume we deal with a homogeneous product family as it was defined in Chapter 1 of Volume 2. In such a case this product family is represented by a line such as (L) in Figure 13.2. At the first thought one would expect that this line could be the straight line OP, point P moving along, when the product size changes, along this line.

This cannot be true for two reasons:

1. Even if the ITF would change linearly, the product structure, due the non-linearity of its algorithm, cannot change linearly,
2. But there is a second reason, more fundamental, because the mix of the ITFs does change, even in an homogeneous product family: a basic observation shows that, when the size does increase, low ITFs increase more rapidly than high ITFs. Let us illustrate that by a simple example: suppose you consider a container for some fluid with a regulation device. This device is composed of a sensor plus a small electronic box activating a tap; in order to quantify its product structure, we analyze this product in terms of its major functions and find:
   - the tank itself plus its support: these are the major parts of the product, at the low ITF levels;
   - the sensor, which uses mainly "positioning" ITF;
   - the electronic box, which at the upper level of the ITF;
   - plus the tap, which is a mix of "containment" plus "positioning".

From this mix, a product structure can be computed. Suppose now that you want to increase the size of the tank in order to store more fluid; the question is now: How the different ITFs are going to grow? Linearly or not? If you have to double its capacity, the size of the tank would have to double, whereas the tap would probably have to slightly increase but the electronic box would remain about the same (except if the tap has to change a lot – but in this case the capacity of the container is not the only thing that is changed as the flow is supposed to change too – and this would imply that the output power of the electronic box must be increased).

This analysis describes a "law of nature": when a product size increases (remaining inside an homogeneous product family), low ITFs increase faster than high ITFs and the difference in change depends on the "distance" between the lowest and the highest ITFs of the product. Practically it means that, starting with a product with a given product structure:

- If the product size decreases, the product structure does increase, as the amount of high ITFs represents a larger proportion of the product.
- If it increases, then the product structure does decrease, as the amount of low ITFs represents now a larger proportion of the product.

Consequently, always in a homogeneous product family, size and product structure move in opposite direction. This can be checked for any family.

This "law of nature" can of course be modelized, and it has been done. It helps explain why, when a "multiplicative" formula (the formula which is a component of a specific model) is used such as:

$$\text{cost} = b_0 \times \text{size}^{b_1}$$

The exponent $b_1$ is always less than 1 if the product size is quantified by its mass (or more than 2 and less than 3 if it is quantified by one dimension only, such as a diameter or a length).

All these facts are consistent and express the same basic idea.

## A Word of Caution

The fact that the exponent $b_1$ is smaller than 1 may give strange results if you break down the hardware into small pieces. Let us illustrate.

Suppose the exponent takes a value of 0.8 – which is an average order of magnitude for hardware – then the cost would go, for a hardware weighting 10 000 units, from 1585 ($10\,000^{0.8}$) if you estimate it without breaking the mass down, to 2512 ($10 \times 1000^{0.8}$) if you break the hardware into 10 pieces. Of course it is always possible to say that the cost of integration is included in the second estimate! But you better check it!

This can be easily described geometrically. Let us start with the basic relationship giving the change of the specific cost (the cost per unit of mass) when the mass changes (Figure 13.3). The shape of the curve is driven by the product structure, its position on the graph by other factors. The curve, as it is displayed, supposes that the product is composed of just one part.

**Figure 13.3** The general relationship between mass and cost.

Let us now investigate what happens if the *same* product A, weighting 100 kg, is now made out of 10 parts, weighting about the same weight of 10 kg per part (Figure 13.4). The cost per kilogram of each part is represented by B.

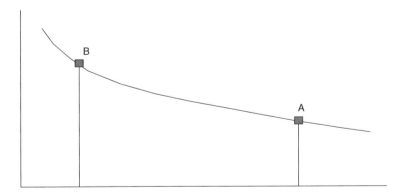

**Figure 13.4** The increase of the specific cost from 100 to 10 kg.

It means that, if we consider the whole product, its cost per kilogram is now represented by C (Figure 13.5).

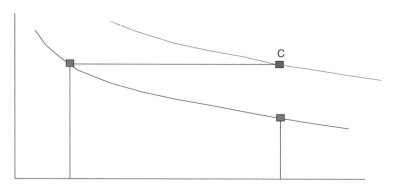

**Figure 13.5** What happens if the product is made of 10 parts of 10 kg each.

The specific cost moves now on a different curve: the product will now cost more per kilogram. Once again this kind of curve can be used to explain to design engineers what happens when they design a product with too many parts.

But this is not the end of the story.

What happens now if the product is always made of 10 parts, with one part weighing 90 kg and the nine other parts 1.1 kg each. The computation now shows that the specific cost is now represented by D (Figure 13.6).

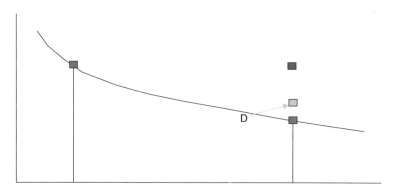

**Figure 13.6** What happens if the products is made of 10 parts of different masses?

This graph shows that the mere number of parts is not enough: the way the parts are distributed must enter into the picture! This is a strong incentive to take into account, when the cost of hardware is estimated, not only the number of parts but also their distribution.

## 13.3 The Quality Level

### 13.3.1 What Do We Mean by Quality Level?

The quality level refers, because we are cost oriented, to the amount of work which has to be done beyond the pure machining activities. Let us consider two opposite level of quality:

1. For consumer products the level of "quality" is very small: this does not mean that the "quality"[5] – as defined in all books dealing with the subject as the capacity of a product to fulfill the needs it was produced for – is low or bad: it simply means that, for a reason which is explained below, practically very little work has to be done beyond the pure machining activities.
2. At the opposite of the scale, what happens for a product realized for a nuclear plant? A lot of activities – all of them having to be recorded (and this has a cost!) – have to take place beyond the pure manufacturing. As an example for a welding operation:
   - the material has to be "certified": traceability up to the high furnace and associated controls;
   - the operators have to be fully certified for the job;
   - the welding machine is tuned;
   - samples are made and tested (and will be saved for future investigation if anything appears in the use of the equipment);
   - the tuning of the machine is controlled;
   - the operation itself is performed;
   - other samples are made;
   - the welding is checked by radiography or any other process.

The right word for the quality level could therefore be here the "quality control effort". But it is too descriptive for our purpose, because assessing the control effort would imply some analytical approach. This kind of approach is not suitable for cost estimating in the early phases of a project (the "decision phases").

## 13.3.2 Quantifying the Quality Level: A Basic Approach

It has now become a tradition in the cost estimators community to assimilate the quality level with the place (the office, the plant, the car or the truck, the ship, the civilian or military airplane, the space manned or not) where the product will have to operate. There is some logic in this tradition, but it is only a starting point, as the quality level, for the same place, may considerably change for reasons which are explained later on in this section.

We proposed a different approach based on the technical analysis of what is required for guaranteeing the level of quality expected by the product customer.

This approach is based on the real cause for improving or relaxing the quality control: Why do we need to control the product? Simply because we want to be sure it will work properly in the environment it will have to operate, for the time it will have to operate, with or without the possibility of maintenance, taking into account the consequences of a failure. The internal operations and the environment impose constraints on the product and we want the product to sustain these constraints: that will be the case if no part of the product works beyond the elastic limit of the material. Maybe this has to be explained.

No material, no welding, no operation generally speaking, can be guarantee default free when it is done. What can be the consequence of a default? This consequence depends first of course on the importance of the default, second on the fact it can be the beginning of a break if the constraints imposed on the material are strong enough.

---

[5] The word "quality" has different meaning. In this section it refers to the amount of work which is done during the manufacturing process for guaranteeing the quality as defined in the text books.

An important concept in this respect is the elastic limit: when a material is subject to an increasing constraint of traction, it extends in an elastic way; this means that if the constraint is removed, the material will find again its original dimension. We say it reacts "elastically". This is true up to a certain limit, called the elastic limit: when the constraint become higher than this elastic limit, the material becomes deformed and does not retrieve its original dimension afterward. This is true if the material has no default. In the presence of a default, when the constraint is increased, the elastic limit in the vicinity of the default is exceeded quicker than in the other parts: the material starts then to break at this place, long before it should in normal circumstances.

Present engineering models allow to compute the constraints in any place of the material of a part and then to design the product in such a way that the elastic limit will never be reached. This assumes that the material is homogeneous. In order to be sure it is the case, quality control must take place. But it is an expensive process.

What can be done in practical circumstances?

- For products which are not mobile but only displaceable – such as a table – the less expensive way is to work far away from the elastic limit, which means that no quality control is necessary (the way we prepare materials is good enough for avoiding major defaults). A table for instance, manufactured for a load of 100 kg, could probably be loaded at 1000 kg; the constraints on a table are probably less than 1/10th of the elastic limit. The price to pay is that the table is heavier than it could be but it is not important because the table is not moved every day.
- For products which are mobile on ground or sea, the weight starts becoming a marketing constraint. In order to get an acceptable weight, materials are more machined in order to get rid of a part of the non-useful material. This means that the material works closer to the elastic limit and that some control becomes necessary.
- For products which are mobile in the air, and *a fortiori* in the space, the weight is a major constraint and the designers try to get rid of all non necessary material. This means that the material works closer and closer to the elastic limit; practically a margin is always added, just to be sure that very little defaults, which could not be detected, will not destabilize the product (an order of magnitude of this margin is 20%). Then a lot of controls are necessary in order to be sure that the elastic limit will not be reached in any point of the material. The products become very expensive.

This classification is based on the mobility of the products, assuming that the consequences of a failure are about the same. If the consequences of a failure can be dramatic (this starts with the aircraft and grows very rapidly for nuclear reactor), the solution is to take a higher margin, sometimes much higher, and to increase the quality control far beyond that should be strictly necessary. This explains the cost of nuclear equipment (of the primary circuit of the reactors for instance).

We believe that this approach of the quality control effort is much more suitable for cost estimating than the conventional approach (which assumes that the quality control is simply driven by the environment in which the product will have to work) for the following reasons:

- First of all this approach "speaks the language" of the design engineers.
- Second it allows for precise adjustment of the quality control effort as a function of the design. For instance design engineers may want to reduce as much as possible the mass of a fuel tank of a satellite (this will allow to carry more fuel

and then to increase the lifetime of the satellite). Using the "standard" quality control which may applies to other equipment is not realistic for this tank which is going to work very close to the elastic limit of the material and for which more quality control must be added. Consequently the cost of this tank will be much higher than could be expected if the "standard" (just taking into account the environment the tank is going to work in) quality control was used.
- Third it allows for preparing trade-offs analysis. Trade-offs analysis is the basic tool for really adjust the design of an equipment or a system to the customer needs. Trade-offs analysis fills the gap between design and these needs. But it requires to make the right balance between mass reduction and quality control. Suppose for instance you want to reduce the cost of a launcher; due to the moderate influence of the first stage mass on the payload mass, efforts should concentrate on the first stage. One way to reduce its cost is to reduce the quality control; in order to keep the same level of reliability, this means that the masses of some parts should be increased in order to have these parts working farther away from their elastic limit. For these parts the cost of the material is going to increase, but as the cost of the quality control will be reduced we can expect to reduce the total cost; in order to check this possibility of cost reduction, we need a way to be able to compute both and consequently get the best compromise between cost and performances.
- Fourth it allows to challenge the effort of quality control to what is strictly necessary due to the design. This means it fosters the dialog between the design engineers and the people who are in charge of the quality insurance.

Consequently the normal way of estimating the influence of the quality control on the product cost must be to establish a relationship between this quality control and the distance the product will work from its elastic limit. This has already been done.

### 13.3.3 A More Complete Approach

However we must go one step further.

In all present general models, what we call "quality level" is, as a matter of fact, a mix of two points. Quality control, when it is associated with the environment in which the product will have to work, means two things:

1. First of all, the level of quality control has to be increased, as described in the previous section, when we want a product work closer to the elastic limit of the materials.
2. But second this variable takes also into account the fact that, in order to get closer to this elastic limit, we have to remove as much as possible material from the slugs the product is made of. This means that the machining time is much more important for this type of product than it is for a product working far away from this elastic limit: this increases the cost a lot.

For "ordinary" products, both things go hand in hand: for moving objects (from surface to space) we want to reduce the mass, which means increasing the quality control. This is the logic on which are built the present general models.

But there are other objects for which both things do not go hand in hand: these are the objects for which the consequences of a failure can be dramatic. We obviously think about nuclear equipments, or more generally for equipments which manipulate dangerous, for the population, chemical products. From the list of the quality controls which are done on these equipments, the level of quality control is nearly similar to space products. But of course there is no reason to reduce the mass, which means that these equipments work away from the elastic limit of the materials.

Consequently, the variable we call "quality level" should sometimes be broken into two parts:

1. One part quantifies the quality control itself.
2. The second part is related to the manufacturing effort which is needed for having the equipments work closer to the elastic limit, when it is needed.

These two parts can be defined as additive or multiplicative.

## 13.4 About the Materials

There is a large variety of material, as Figure 13.7 illustrates very briefly.

| Material | MJ/kg |
|---|---|
| Cast iron | 17.0 |
| Steel | 20.9 |
| Sheet steel | 27.4 |
| Iron for concrete | 29.5 |
| Steel tube no welding | 32.2 |
| Wire | 30.4 |
| Copper | 13.6 |
| Zinc | 45.1 |
| Lead | 10.0 |
| Nickel | 37.4 |
| Molybdenum | 334.4 |
| Chrome | 167.2 |
| Tungsten | 292.6 |
| Aluminium | 214.6 |
| Magnesium | 371.1 |
| Cement | 4.8 |
| Plaster | 2.3 |
| Bricks | 3.1 |
| Reinforced concrete | 4.3 |
| Glass | 25.9 |
| Polyethylene | 88.6 |
| Polystyrene | 128.3 |
| Polypropylene | 100.1 |
| Synthetic material (cloths) | 248.5 |
| Paper | 19.2 |

**Figure 13.7** Energetic content of industrial products in MJ/kg.

The cost of these materials differs due to the amount of energy needed to produce them from mining operations to the finished product (the cost, as you would certainly guess, also depends on the production quantity).

There are two elements to take into account when using a material: these are the procurement cost and the machining cost.

One idea to do it can be split into three steps:

1. A reference material is selected: one generally of course chooses the most common material for his/her industry type. If any tuning of the model is necessary, it will be done on this material.
2. The procurement cost of the new material relative to the reference material is then entered in the model.
3. The difficulty to machine this new material, always compared to the reference material, is then also entered. Fortunately enough tables have been prepared long time ago by the mechanical engineers: these tables were built by selecting a material to which a "machinability" index of 100 was given. Then other materials were studied and their more or less higher difficulty to machine them – compared to the selected reference material – was measured; an index was computed giving in fact the easiness (the larger the index, the easier is the material to be machined). Figure 13.8 lists some published values, as examples.

|  | Metal | Index |
|---|---|---|
| Carbon and free machining steel | Bar 1020<br>Sheet and strip 1095<br>Steel bar | 90<br>80<br>100 |
| Alloy steel | Bar 4130<br>Casting, Sand 4230<br>Casting 4320<br>Forging 4330<br>Plus vanadium 4340<br>Tool steel hot work | 60<br>60<br>60<br>40<br>15<br>37 |
| Stainless steel | Sheet 301<br>Bar 303<br>Bar 430<br>Bar 440C | 50<br>70<br>60<br>40 |
| Titanium alloy | Bar | 30 |
| Iron base alloys | Armco<br>Bar A286 | 25<br>15 |
| Cast iron | Ductile<br>Gray | 100<br>100 |
| Nickel base alloys | Astrolot<br>Hastelloy<br>Inconel | 15<br>10<br>25 |
| Aluminium alloys | Casting 108<br>2024<br>5052 | 180<br>150<br>190 |
| Magnesium | All forms | 500 |

**Figure 13.8** Illustration of the machinability index.

### 13.4.1 About the Machinability of Materials

Machinability of materials has become a parameter often used[6] by cost analysts who have to use different materials. It therefore deserves a few words of explanation.

The ASTE (American Society of Tool Engineers) recognizes[7] that the term is difficult to defined exactly. It explains that the easiness which a given material may be worked depends on several variables:

- common machine variables (cutting speed, tool material, etc.);
- common work material variables (hardness, etc.);

and concludes that three quantities must be taken into account to quantify this ease: tool life, tool forces and power consumption, plus finish.

It then discusses on how to measure these three quantities and introduces, in relation with the first one, the "machinability rating" as the relative *speed* to use with a given material to obtain a given tool life. A reference material (AISI B1112 steel) was selected and given a machinability value of 100. A machinability rating of 50 for a material[8] means therefore that this material should be machined at half the speed of the reference material in order to get the same tool life. This explains that machinabilities are higher than 100 for soft material such as aluminum, and lower than 100 for hard materials.

Machinability consequently represents relative cutting speed; it is then understandable that it has an inverse relationship with cost.

The ASTE recognizes that the machinability bears a rough relation to the Brinell hardness and the tensile properties.

Machinability refers only to machining operations. If welding is considered, a "weldability" should be defined,[9] as well as a "bendability" for bending operations, etc.

## 13.5    The Manufacturing Process

The manufacturing process can be a very important cost driver, at least for mechanical products. For instance for making a cylinder, you may use a casting process, or a forming process, or a forging process, or any other one.

Quite obviously the manufacturing cost depends on the process that it to be used. This needs another variable to take it into account ...

## 13.6    The Basic Relationship Between Size and Manufacturing Cost

Two fundamental subjects must be dealt with: the type of formula to be used on one hand, the fact that production may differ largely in different industrial sectors.

---

[6] Even for the preparation of cost-estimating relationship (CER) when the cost of products made out of different materials has to be prepared.
[7] Tools Engineers Handbook, McGraw-Hill Book Company, 1949, p. 319.
[8] Qualitative ratings (A, B, C and D) are also sometimes used: A indicates a high cutting speed; D, a lower speed.
[9] It is generally given as a letter, from A (excellent) to D (poor).

### 13.6.1 The Type of Formula

The multiplicative formula:
$$\text{cost} = b_0 \times \text{size}^{b_1}$$
proved itself good as soon as the product size becomes important. It can be kept, taking into account the fact that the cost estimator should check, when this size becomes large, if a change of technology does not occur.

However the "correction by constant" formula:
$$\text{cost} = a + b_0 \times \text{size}^{b_1}$$

must be preferred when the size becomes small, at least when the quantity to be produced is small. If it is large the cost of preparing the work and tuning the machine becomes very small if all products are manufactured simultaneously: as usual no solution is perfect all the times!

### 13.6.2 The Type of Production

A simple look at the way companies are organized for production convinces us that there are several types of production; among these types one can recognize three basic types which may be defined as follows.

*Small Quantity Production*

Small quantity production is generally done on "universal" machine tools. The company which does it has a set of such machines and programs the production in order to get the most efficient use of them. It therefore masters the schedule.

This is the classical production type. In this type one can observe the learning effect, such as discussed in Chapter 8. Consequently the average cost per product decreases when the quantity increases.

This production type is often called "no constraint production" because the factory manager is relatively free to organize the production, as long as of course he/she meets the clients' demands.

*Large Quantity Production*

Now the production is done on dedicated machine tools. Robotization is then an important factor. Quite often a plant is built for this type of production only.

This production is described in details in Chapter 9. Notice that no learning is mentioned. This situation comes from the fact that the process is nearly rigid, once it has been decided (think for instance to the assembly line of cars): the work division is complete, each operation has a predefined time to be done and nothing can easily be changed. The cost is driven by the plant capacity, and certainly not by the quantity. Therefore the parameter "plant capacity" is the important one, plus the load as it is defined in Chapter 9.

This production type is often called "constraint by the plant capacity". Nevertheless the manager decides on the capacity, and its change, and therefore on the schedule.

### Intermediate Production

This is an intermediate case where the customer decides about the production rate, and may change it from time to time; the producer has then to adapt himself to these changes.

This situation is quite frequent when there is only one customer and does happen in the contracts between a company and the government.

The company has a set of semi-automated machine tools, because it needs flexibility. Flexibility is the key word for this type of production; two flexibilities must be considered: for an increase in the production rate and for a decrease (as we saw it in Chapter 9, operations are not symmetrical).

Decreasing the production rate always means an extra cost, for reason already mentioned in the dedicated chapter and because the laws, at least in Europe, do not allow a fast reaction. After some time – and the duration of the adaptation phase is also a parameter – production stabilizes, but always at a cost higher than before.

All that can be modelized, and has been.

## 13.7 Adjusting the Model to the Manufacturer: The Concept of Industrial Culture

This is the important variable which allows to make estimates without a reference product (it can be considered as a substitute of such a reference). This section will deal first with the culture in its general meaning before it can be apply to organization.

The word was coined by anthropologists. A general definition of the culture was first given by E.B. Tylor:[10] "Culture ... is that complex whole which includes knowledge, belief, art, morals, law, custom, and any other capabilities and habits acquired by man as a member of society". Obviously that should influence the way we work!

Kluckhohn and Kelly[11] wrote: "A culture is a historically derived system of explicit and implicit designs for living, which tends to be shared by all or specially designated members of a group". Some words are worth an emphasis:

- *"Historically"*: culture develops itself with time; it does not emerge in a vacuum, but by the interrelationships of individuals who, slowly, but surely, develop it.
- *"System"*: culture is a very complex set of statements which must considered as a whole, even if we have, to get a better understanding, to analyze its components.
- *"Designs for living"*: Generally speaking culture is defined as *"the way people feel, think and act"*. Sociologists prefer to put the emphasis on the living because this is what they observe, but the culture is mainly in our heads.
- *"Shared"*: each individual has his/her own way of acting and this makes his/her personality. The culture is what we have in common.

---

[10] E.B. Tylor. *Primitive Culture* in Leslie A. White, *The Concept of Culture*, American Anthropologist, April 1959.
[11] C. Kluckhohn and W.H. Kelly. *The Concept of Culture*, in R. Linton (ed.). *"The Science of Man in the World Crisis"*.

This general definition of culture immediately appears when we go from one country to another one: people, in their general behavior, do not think and act the same way as other people.

When the British say: "we are different", this is exactly what they mean, and all the inhabitants of a given country, and often part of a country, could say the same thing.

George C. Homans who studied[12] in depth the behavior of the small group (defined as a set of people who frequently and directly *interact*: a group is created by the direct interactions of its members) distinguishes the "external system" and the "internal system" (this will also be applied to the organization). The *external system* is the group's environment; this environment has its own culture: to give a general example, the behavior of the same group immerged in India would be different as if it were immerged in France. This is generally what we mean by "culture" when we speak about it without referring to a particular organization. George C. Homans uses three terms to describe the behavior of a group: sentiment (what the people feel about each other), activity (what they do) and interaction (how often and why they interact); he adds that the behavior of a group, as observed from the exterior of the group (for instance its productivity, the quality of the job done, etc.), must be such as to allow it to survive in the environment.[13]

The *"internal system"* is the set of stable relations which develop inside a group. How does this set of relationships develop? Let us read what Chester Barnard[14] wrote about this question (in 1938: the subject is not recent!): "when the individual has become associated with a cooperative enterprise he has accepted a position of contact with others similarly associated. From this contact there must arise interactions between these persons individually, and these interactions are social. It may be, and often is, true that these interactions are not a purpose or object either of the cooperative systems or of the individuals participating in them. They nevertheless cannot be avoided ... Hence, cooperation compels changes in the motives of individuals which otherwise would not take place". Homans defined the internal system as the elaboration of group behavior that simultaneously arises out of the external system and reacts upon it.

He adds what could be told as the central theorem about the elaboration of the culture: "the more frequently persons interact with one another, the more alike in some respects both their activities and their sentiments tend to become". A central concept to describe the group's culture is the concept of *norm*: "a norm is an idea in the minds of the members of a group, an idea that can be put in the form of a statement specifying what the members should do, ought to do, are expected to do, under given circumstances".

We are back to "the way people feel, think and act"! We are now equipped to study the organizational culture.

The *"organizational culture"* will be one of the leading concepts of this section (a second one will be the "need" as described below).

The organizational culture is immerged in a more general culture, in which we can distinguish two levels. First of all there is the culture of the whole country: Americans all share some norms and values (and most of them are also shared by a more general culture (the western culture); there is no need to emphasize this statement.

---

[12] George C. Homans. *The Human Group*. Routledge & Kegan Paul Ltd, 1968. We strongly recommend you read this book, even if it is not recent. It gives in depth perspective on how the culture of a group emerges.
[13] Exactly the same thing can be said about the organization as a whole (observed from its exterior), or about a facility (observed from the organization, but externally from the facility).
[14] Chester Irving Barnard, *The Functions of the Executive*. Harvard University Press, Cambridge, MA, 1938.

Secondly there is the culture in the industry the company works for and receives orders from. We are here only interested by this culture, which we will refer to as the "macro-culture". This is what Homans calls the "external system".

Inside this macro-culture, the company develops its own organizational culture, which borrows many norms from the macro-culture it is immerged in (to allow it to survive in its environment), but also has its own norms which develops from the fact that its members generally have more interactions between themselves than with people of the environment (people who spent much of their time outside the organization have often difficulties to integrate themselves inside the organization: they had no time to acquire its own norms): this is Homans' internal system. We will refer to it, from now on, as the culture.

Inside the company, "subcultures" develop, as Homans described the process. But we are not interested at this stage in those subcultures.

So the culture is the framework inside which organizations operate – or do business, if you prefer this formulation. Most often it is unconscious (and is expressed by sentences such as: "we have always done things this way!", which means exactly the same thing) and appears when you go from one company to another company (they do things differently!), not always inside the same type of industry but mainly when you go from an industrial sector to another one: culture is generally not associated with one company but with a whole industrial sector what we call the macro-culture. From the preceding introduction, we can say that two reasons explain this fact:

1. First of all engineers have much more contact with people inside the same type of industry than with others.
2. Engineers generally move from one organization to another one inside the same sector. John M. Pfiffner and Frank P. Sherwood wrote:[15] "Professionalization is the consequence of a highly skilled and specialized society. The twentieth century in the United States has seen the rise of host of new professions and expansion of others. As a mean of upgrading performance in these various efforts, much emphasis has been placed on standards of conduct that relate to the profession and not to the institution".

The culture then trends to homogenize inside a whole industrial sector. So, the "culture" impregnates the way engineers "think" about products, the way they design products (this is one of the most important aspect of the culture), the type of machines which are bought, the organizational structure, the effort devoted to cost, or even more petty things such as the amount of time spent in meetings, or the way meetings are prepared and managed, etc.

How can we compare organizational cultures? This is very difficult, especially if we want to quantify the cultural differences: it is easier to quantify the results than the causes (but such a quantification of the culture has already been done, which is valid only in the domain of cost). On the surface – because this analysis cannot take into account for instance the way engineers think about their products – we suggest to study the article of Luther Gulick,[16] quoted by Edmund P. Learned and Audrey T. Sproat, who distinguishes 11 rubrics: division of labor or specialization, departmentalization on the basis of purpose, process, clientele, or place, coordination

---

[15] John M. Pfiffner and Frank P. Sherwood. *Administrative Organization*. Prentice-Hall, Inc., Englewood Cliffs, NJ, 1960.
[16] Luther Gulick. *Notes on the Theory of Organization*, in Luther Gulick and Lyndall F. Urwick (eds). *Papers on the Science of Administration*. New York, Institute of Public Administration, Columbia University, 1937.

by hierarchy, ideas or committees, decentralization, unity of command, staff and line structure, delegation, span of control and leadership style. Regarding this last rubric, Joan Woodwart[17] wrote: "It appeared that the technical methods were the most important factor in determining organizational structure and in setting the tone of human relationships inside the firm". We would like to add, from our own experiences, the presence or the lack of competition.

One thing which is remarkable about the organizational culture is its *stability*. An organization can be seen as a set of people *who indoctrinate any new member in order to have her/him share their culture*. As one cost estimator said: "in order to change the culture of an organization, you must move the personnel in a different plant, demolish the factory, rebuild it in a remote place and hire new personnel". There is some truth in this statement: the culture is really part of the organization. So, do not believe it is easy to change a culture; not being a sociologist, we cannot really advise you how to change the culture of an organization and it is not the subject of this section. But we know it takes a lot of time and effort for the managers and the personnel; we can only tell, from past experience, that everybody, from top management to design engineers, has to be trained at the same time or the training is useless: a culture is, as the preceding discussion showed it, something we share with others; a culture is a background of whatever we do or say every day. The consequence of that is: use any quantification very carefully; this quantification tells you what would happened if the culture were changed, it does not say "you just have to change the culture in order to observe these modifications".

Many sociologists have already studied what they call the "*resistance to change*" and they developed interesting ideas for reducing this resistance. This resistance is not always described by the people concerned as such: they use different words to express the idea and managers have to be trained to understand the concepts behind the words. In any case, the help from sociologists might be welcomed.

The culture must not be confused with the level of quality control, which is of a different nature and is, as we saw it, a technical input which depends on the way products are designed, and then obviously depends on the culture, but is a consequence, not a part, of the culture. Some people often associate "level of quality control" (How much control do we do? How many people are devoted to this task? How much do they cost?) with "culture" but we think they are different concepts.

A closely related concept to culture is what has been studied – for instance by SSCAG (Space Systems Cost Analysis Group) – as a "*new way of doing business*". It expresses the same idea, with less strength: a new way of doing business will change "the surface" of the things (we should do less of that and that), without changing, for instance, the way engineers think about their products. This cannot be changed so easily and requires something different; but the cost comes primarily from that: someone already said that cost takes its source in the design, but we would prefer to say cost takes its source in the way engineers think about the product, that is to say from the organizational culture. Nevertheless the concept is interesting and deserves some attention. Tables 13.1 gives some reduction factors attributed to the "new way of doing business"[18] for development and production.

---

[17] Joan Woodwart. *Management and Technology*, London, Her Majesty's Stationery Office, 1958.
[18] *Source*: Information from SSCAG (the "Space System Cost Analysis Group") hardware survey, January 1998, to allow for: reduced testing, use of IPTs (Integrated Product Teams), limited changes, automated production.

**Table 13.1** Reduction factors attributed to the "new way of doing business" for development and production.

| Activity | Development | | | |
| --- | --- | --- | --- | --- |
| | Engineering labor | Tooling labor | Model material | Total |
| Concurrent engineering | 0.88 | 0.91 | 0.93 | 0.81 |
| Use of IPTs | 0.89 | 0.94 | 0.95 | 0.84 |
| Rapid prototyping | 0.88 | 0.75 | 0.96 | 0.82 |
| Use of commercial specifications | 0.88 | 0.90 | 0.85 | 0.76 |
| Others (?) | 0.91 | 1.00 | 0.90 | 0.91 |
| Overall | 0.85 | 0.82 | 0.90 | **0.72** |

| Activity | Production | | | |
| --- | --- | --- | --- | --- |
| | Engineering labor | Production labor | Flight model material | Total |
| Concurrent engineering | 0.81 | 0.82 | 0.95 | 0.74 |
| Use of IPTs | 0.90 | 0.91 | 0.96 | 0.85 |
| Rapid prototyping | 0.91 | 0.83 | 0.94 | 0.78 |
| Use of commercial specifications | 0.89 | 0.86 | 0.90 | 0.74 |
| Others (?) | 0.90 | 0.85 | 0.90 | 0.82 |
| Overall | 0.82 | 0.83 | 0.90 | **0.73** |

It has been said the "new way of doing business" produced a factor of 0.66 on the production of the Crusader.

### 13.7.1 How the Culture Can be Used in a Model?

The first thing to do is to establish a scale; to start with the level of quality control can be used as a guide, because the company which share the same level of quality control generally have more interactions between themselves than with other companies. Once this is done it can be refined.

Any scale can be used, the important point is that it has to be logic.

Once this is done, the culture becomes a variable as the other ones and can be added into the model algorithms: it adjusts the cost to the companies way of doing business.

### 13.7.2 Conclusion

The culture is what was missing to take into account the cost differences between the companies.

This concept has of course limitations, the most important one being it requires a good level of competition: companies in the same business line share the same culture not because they like to do that, but because competition forces them to do so.

This means that it is probably not well adapted to monopolies. In such a case using a reference product is necessary. However there is no guarantee that a monopolistic company will follow the logic of cost formula if it has not do to it! Cost estimating a product made by a monopolistic company is a difficult point.

## 13.8 A Last Word

This chapter shows that general modelization is not finished: a lot of improvements can still be done.

In the course of this chapter some improvements are mentioned. We hope it will foster comments and further improvements.

# 14 Modelization in Other Classes

## Summary

In this chapter, one shows that universal models are not only dedicated to hardware products, even if the hardware may be still the major domain of application.

This chapter briefly gives some information about some other domains of applications. Several other domains could certainly benefit from the concepts.

In Chapter 13 we were only concerned by hardware, and for this class of products by production. This chapter illustrates the concepts already mentioned to different class of products. As an introduction to the subject, the presentation is limited to the software, the buildings and the tunnels. The reader will observe that the basic concepts are similar, even if some modifications must be taken into account.

The presentation is strictly limited to the basic concepts.

## 14.1 Software

The software industry does not escape the rule; in order to make an estimate, you need three types of basic information:

1. its size,
2. its intrinsic difficulty,
3. its environment of development.

### 14.1.1 About the Size

The software size is, according to all the cost estimators, the major problem to be solved in the early stages of a project. A lot of effort, which is impossible to describe here, has been devoted to this subject, the first one being the definition of a metric (What do we measure and how?). Three directions were investigated for quantifying the size:

1. *Using the physical number[1] of lines of code.* This solution is very similar to the use of mass for describing the size of hardware; however if the mass of hardware has

---
[1] Not taking into account the lines of codes automatically generated by the compiler.

in principle one definition only, the definition of a "line of code" or "statement" or "instruction" is far from obvious: Do we include the comments? the job control language? Can a developer write the same statement with just one or several lines of codes? ... The Software Engineering Institute (SEI) published a framework for this analysis, framework which can be a good start for the beginner. An interesting question about the size is the nature of language: the more "developed" the language (one generally speaks of the "order" of the language), the less the number of lines of code are required for performing a given function. This question is rather similar to the question of the material for hardware. We saw in the previous chapter one solution which can be used, solution which tries to use, for a given material, adjustment made to the model for another material. The situation is different for software; in fact two questions may be asked:

(i) Suppose the software size is known in a given language: What could be its size in another language? This can be solved by using ratios. These ratios only give an order of magnitude.

(ii) Does the language has an influence on the cost, supposing of course that everything else (experience, tools, ...) is similar? This question is controversial, some authors – such as Barry Boehm[2] who wrote "the amount of effort per source statement is highly independent of language level" – saying that the language has no influence on the costs, whereas others saying it does. The language has an obvious effect on the software size: this is the main purpose for using a high-order language. Therefore fulfilling the same function with two different languages may produce different costs, even if the cost of writing an instruction is the same; therefore one should pay attention when using the term "productivity": productivity per function may increase when changing the language, whereas productivity per statement does not.

2. *Using "function points"*. When created (Albrecht, 1975) the purpose of function points was to measure performance of developers. Function points are related to "What is to be programmed (the functions)?" and not "How is it to be programmed (the statements)?". This explains why the first version of the function points analysis used information about the number of external user inputs, the enquiries, etc., which are oriented toward business management software applications. Function points knew through the years a lot of refinements;[3] one of them was published [55] under a different name.

3. Another approach was tried in order to go directly *from requirements to size*. It could be a very good idea if it were possible to get a generally accepted method for writing the specifications.

Returning to the size quantified from the number of source statements, it is obvious that this size can be estimated the same way as costs are: analogy, expert opinion, etc. Models can also be built for this purpose and, as usual, one can distinguished between specific and general models.

Specific models use correlations between the software size and the environment in which the software must work; for instance it is possible to correlate software size with the number of users, the number of printers, etc. Such models are developed in the usual way, ... as soon as you get the data.

---

[2] Barry Boehm nevertheless had to make modifications to his model for Ada.
[3] There is an international function points user group (IFPUG) where documentation can be found about these function points.

An interesting approach to general modelization was proposed by Georges Bozoki in 1993 [13]. This approach, oriented toward size estimating at the proposal stage, is based on the following key facts:

- Qualitative sizing information is more accurate than quantitative one.
- The human mind is more comfortable with estimating relative sizes of modules than absolute sizes.
- Estimated and actual relative sizes are strongly correlated.

Therefore the method starts by a breakdown of the software in terms of modules and asks the expert to provide pair-wise comparisons between these modules: this is typically a subjective approach which can be checked by one method such as the one mentioned in Chapter 6. The result is a hierarchy between the modules; if the size of two modules can be directly quantified, then a statistical procedure may determine the size of all of them.

Once the problem of the size is solved, it is time to go from it to the cost of the software. This implies to take into account other considerations.

## 14.1.2 Relationships Between Cost and Size

As usual building a cost model may start:

- either by assuming a relationship between cost and size,
- or by building such a relationship by some statistical analysis of existing data, other things being taken into account afterwards in order to reduce the residuals.

As models developers do not – with one exception – disclose their relationships, we will limit the discussion here to the second point, which is a major concern for the cost estimator.

Barry Boehm [8] in his book mentions several relationships – made by different people – which all use a simple multiplicative formula. The interesting thing is that the exponent on the number of instructions goes from 0.91 to 1.83, a very large range indeed! Barry Boehm himself proposed three similar relationships (based on the level of constraints which must be taken into account by the developers) with an exponent (the exponent has to be applied on lines of code given in thousands) going from 1.05 to 1.2. Does that make a large difference? For a software of 100 000 lines of cost, the cost ratio is about 2; for a software of 10 000 lines of codes, the ratio becomes 1.4: the influence on the cost of the constraints is not negligible.

The model then presents three sets of algorithms. These three sets cannot be considered as using a qualitative parameter because the way they are built conflicts with our definition of qualitative variables, as it is given in Chapter 5 of Volume 2: the formulae use different slopes depending on the qualitative variable. This means that Barry Boehm decided to distinguish three subsets in the whole set of software.

One thing seems to be well established in the software industry: when using a multiplicative formula, the exponent is larger than 1 and this makes a major difference between hardware and software. This seems to result from two major causes:

1. The larger the software, the larger the number of people who are working on the project, and therefore the larger the communication needs. Communication problems were solved long time ago in the hardware industry through drawings

and prototyping; software development seems not to have solved this problem, but things are improving.
2. Software parts are often largely dependent on the other parts: a small change in one part (for instance in the data base on which computations are built) may have consequences in many other parts.

However it appears that there is a secular trend toward a decrease of this exponent, due to a better software project management; in particular the trend to shift the peak of activity toward the design.

## A Word of Caution

If you use cost algorithms with an exponent higher than 1, the cost depends on the way you structure the software. Suppose you have a software of which total size is 10 000 lines of code:

- If you estimate it directly you get a cost proportional to 63 095 units ($10\,000^{1.2}$).
- If you split it in 10 modules of equal size, you get a cost proportional to 39 811 ($10 \times 1000^{1.2}$).

You therefore get a ratio of about 1.6, which is quite large! It is of course always possible to say that the cost of the 10 000 lines of code includes the cost of the integration of the modules, whereas the cost of the 10 modules does not. In such a case, you should check if 23 284 units is a realistic figure for this integration: maybe it is, maybe it is not! This is a strong incentive to always break your software down to *modules of similar sizes* and to add the integration cost. Your results for cost estimating different software will be more consistent.

### 14.1.3 Taking into Account the Other Cost Drivers

COCOMO is, at our knowledge, the only model which has been fully published.[4] The cost analyst should therefore certainly study it and learn lessons from it. Our purpose here is not to describe it, but to look at its structure in order to help the cost analyst who decides to understand the way it was built. COCOMO – which stands for "COnstructive COst MOdel" – is a cost model for software developed by Barry Boehm and completely published in his book in 1981 [8].

The framework of the model is to split software development into "phases" and "activities". Phases can be viewed as a time breakdown of software development: phases are defined by the definition of endpoints. Inside a phase the work is broken down into activities. The fact that phases and activities bear most often the same name is a bit confusing for the beginner, but quite understandable. COCOMO computes the cost as a whole, and then distributes it on the phases and activities according to ratios, ratios which depends – slightly – on the software size.

Barry Boehm distinguishes 15 other cost drivers beyond the size grouped in four sets of attributes (product, computer, personnel and project). Each cost driver is measured by a subjective parameter, where four to six peg-points are mentioned: very low, low, nominal, high, very high and extra high. A few hints are given to position

---

[4] There are a lot of commercial models for software cost estimating.

the parameter inside this set. Then to each peg-point is attributed a figure between – at a maximum – 0.7 and 1.95, called "level of effort multiplier". The product of these figures gives a global multiplier to the "nominal cost".

*Software Maturity: An Interesting Approach to the Organizational Culture*

The concept of process maturity (due in particular to the work of Watts Humphrey and his colleagues at the SEI) is very close to the concept of organizational culture which was the object of a section of Chapter 13. To understand it, we reproduce here some sentences written by Edward Yourdon.[5] Five levels exist in the SEI model. Let us see how E. Yourdon defines them:

1. *The initial level*: "anarchy prevails – programmers consider themselves to be creative artists, not subject to rules or common procedures"…
2. *The repeatable level*: "organization is characterized by tribal folklore … Another interesting *cultural* distinction: while the success of the level 1 organization depends on the skill of its bottom-level technicians, the success of projects within the level 2 organization typically depends on the skill of the project manager"…
3. *The defined level*: "the key point about level 3 organization is that the process has been codified and institutionalized; everyone in the organization can point to the 'bible' and say 'this is the way we do things around here' " …
4. *The managed level*: "the level where organization is measuring the process and therefore improves it".
5. *The optimizing level*: "the level where there is 'continuous, ongoing process improvement, based on the metrics already captured in level 4' ".

Doesn't all these definitions look familiar? Don't they reflect the way people feel, think and act? The maturity model does not insist so much about tools or metrics, but on the way people think and act: it really speaks about the organizational culture.

One thing which is added by Yourdon: it takes time to move from one level to the next one. He insists: "it is reasonable to expect a period of 2 to 3 years between levels, that is a decade to go from level 1 to 5!". We find again here the question about the stability of the culture.

## 14.2 Building

The construction industry is one of the most important one in any country, but to our knowledge, not so many models do exist in the commercial market for what we call the **decision phase**. Traditionally building cost is then estimated using a ratio (so many € per square meter); later on, during the validation phase, cost estimates are made at a very low level, always using ratios. However some manuals, even for the decision phase, do consider that the ratio changes with the surface as illustrated in Figure 14.1. This figure is interesting as it shows that the change is not linear at all. It could easily be given as a formula; such a formula should include IF … THEN … ELSE statement.

---

[5] In his book *"Decline and Fall of the American Programmer"*. Yourdon Press. 1993.

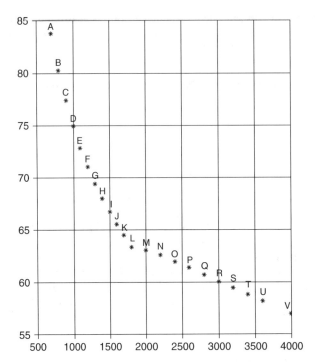

**Figure 14.1** Cost per square foot as a function of the surface in square feet. *Source: National Building Cost Manual.* Craftsman Book Company, 1996.

Nevertheless some interesting comments[6] can be done about this industry.

The logic of a general model for buildings follows the same principles discovered for hardware: the building to be realized has to be described, and a "reference" building selected. However the cost of this reference building has to be "cleaned" from all the special considerations (feasibility factors, quality level, localization factors) before it can be used as a basis for costing the new building. Then this basic cost will have to be adjusted to take into account its special considerations. This makes the creation of a general model for building more complex, but nevertheless feasible; the main advantage of this procedure is that any building can be taken as a reference.

One point has to be mentioned here. We said about hardware that "artistic" products escape the field of our general models. This is also true for construction: special, original, features cannot easily be taken into account. Of course if the reference building already has these special architectural features, the problem is solved. For other ones what the model can do is to estimate the cost of the new building as if it were "standard" (which means similar, from an architectural point of view, to the reference building); then a qualitative variable should be introduced to take into account the original features.

---

[6] Most of these comments are taken out from the documentation of the INSITE model (commercialized by Prime Time, Pennsylvania). This model is also able to cost estimate the site improvement, at the conceptual phase: cleaning, roads, utility supplies, fences, parking, etc.

We are used, in the hardware cost estimating community, to the decomposition of the hardware according to the product structure (system, sub-systems, equipments, modules, …). The same decomposition is used for building construction: total, structure, exterior, interior; at their turn these levels may be broken down: for the exterior, we have the roof and the external walls; for the interior, the partitions, the floors and the ceilings). This makes the architectural original features easier to handle, because only some parts of this breakdown are concerned by them. The "reference" building mentioned upwards may change from one level to the other level.

Another feature which has to be taken into account is the simultaneous construction of several similar buildings in the same area. This decreases the cost of these buildings, about (but not exactly similar) the same way as the "learning effect" described for hardware in Chapter 8.

In this construction industry the description of a building also needs three major concepts: the size, the feasibility and the quality level.

### 14.2.1 The Building Size

As expected, the building size is generally quantified by the floor surface.

However for special building (conference room, auditorium, churches, …) the volume is more important than the floor surface for describing the building size.

### 14.2.2 Feasibility Factors

Two kinds of buildings may have to be considered when describing the feasibility factors.

For "industrial" buildings, the load the surface will have to carry is certainly one important cost driver.

For "ordinary" buildings (including parts of industrial buildings), several feasibility factors have to be taken into account:

- the number of floors,
- the destination of the building, from offices to dormitories, from industrial to commercial, from store rooms to computer rooms, etc. As a building floor may have several usages, the portion of each has to be given.

### 14.2.3 The Quality Level

The quality level is driven by two considerations:

1. Special destination of some rooms of the buildings. Some rooms may have special destinations for which a special quality level is required: this is the case for rooms in which chemical or radioactive products may have to be manipulated or stored.
2. The environment in which the building has to operate. We refer here to two aspects which may have an important influence on the building cost:
    (i) The *seismicity* of the ground on which the building is constructed. There are several levels of seismicity which are well known (from 1 to X on the

Mercalli scale); the special effort to be taken into account depends also on the destination of the building and mainly on its occupancy.[7]

(ii) The *wind and tornado* level. As the wind has no effect on the underground part of the building, the part of the building which is under the ground surface has to be specified.

### 14.2.4 Localization Factors

It is well known that the localization of a building has an important influence on its costs. Two causes do influence the building cost:

1. The *climate*, from tropical to arctic. This has become in recent years an important factor, linked to the Kyoto protocol: in order to reduce the power consumption, building has to be better isolated than it was before.
2. The *place* where the building is constructed. The fact that the cost depends on the area is well known and published (the cost depends on the resources availability, and on the general salary level of the personnel in the area). But there are other causes which decreases the productivity level: the distance at which the building is located from residential areas, the security level and the location (a building created in an empty place is less expensive than a building realized in the middle of a busy city).

## 14.3 Tunnels

A general model is also possible[8] for cost estimating the realization of tunnels, underground rooms, shafts, etc. Such a model is an important need during the decision phase of a project because a tunnel represents an extremely high cost and that several alternatives are always possible. The investor must therefore be able to select the less expensive one, during the decision phase.

The second reason is to be able to answer to the "What if?" question. The exact nature of the rock – and the cost of excavating a tunnel depends a lot on this nature – is practically impossible to know for the whole length of the tunnel, even if preliminary small galleries have been excavated for getting a sample of the rock. But the investor is never sure if the rock will not be inhomogeneous or if a large flow of water may not happen at some place(s). It is therefore important to be able to cost estimate what is going to happen if such or such a problem is encountered during the excavation. One can say that the "What if?" question are more important for this class of "product" than for any other class.

The logic for building a model is always the same. But we present briefly the way a cost model for tunnels can be done because it has special features.

---

[7] See for instance, in the USA, the *"Uniform Building Code"*, published by the International Conference of Building Officials.
[8] Most of these comments are taken out from the documentation of the SubCOST model (created and commercialized by 3f, France).

### 14.3.1 About the Size

For a tunnel the obvious size quantification is the section and the length of the tunnel. Some cost estimators may use the volume to be excavated, but the length of the tunnel is an important parameter for two main reasons: the fix cost (for ventilation and conveying the excavated rock) depends on it, the variable cost also depends on it as it does not cost the same amount, for an identical volume, to excavate a large tunnel with a small length or a small tunnel on a great length.

Another element is the shape of the section of the tunnel because it may drive the machinery which is to be used.

For cavities the dimensions are also better than the mere volume because excavating may be done in several steps, which depends on its height.

The same reasoning may be done for shafts.

### 14.3.2 About the Rock

The type of rock is, after the tunnel size, the most important cost driver. Several parameters are necessary for being able to estimate the cost, the first one being the compressive strength of the rock, the second ones being related to its structure.

#### The Compressive Strength

The compressive strength is to the rock what the machinability is for materials. It is given in MPa (or megaPascal); for instance, the granite is characterized by 180 MPa.

The interesting point – from a cost analyst point of view – is that the excavation cost does not grow linearly with the compressive strength: as a matter fact it grows linearly up to about 100 MPa, and then with a different curvature [11] for higher values. Fortunately enough the change of slope is at about the same compressive strength for all rocks.

#### The Rock Structure

The rock structure has to take into account several features: the rock "quality" (the importance of the fracturing), the discontinuities, the joints nature and the water inflow.

### 14.3.3 The Excavating Technique

A cost model must take into account for excavating tunnels the technique to be used, from explosives to tunnel boring machine (TBM).

When a TBM is to be used, important elements of the costs are the depreciation of the machine and its installation/deinstallation.

### 14.3.4 Several Costs

Several costs have to be estimated for making a tunnel:
- The boring itself, including the fix and proportional costs.
- The mucking (which depends on the distance).
- The linings: several linings are possible.
- The temporary and permanent supports, from sprayed concrete, to bolts and arch ribs.

As usual, studies and management costs must be added.

The purpose of these sections is that the same techniques can be applied to different sectors, but also that no general rule can be given: each sector has its own specificity which has to be discovered by the cost analyst.

# 15 A Word About the Future

In most sciences specific and general models coexist, an interesting thing being that specific models can be deduced from general models, as a simplification of them.

This is not the case in the cost domain where specific and general models live their own life independently.

We believe that general models are still in their infancy. As long as we will not be able to infer specific models from general ones, it will show that we do not still master the process!

We believe that the solution will come from a complete change of our point of view. As one person once said: "Edison did not invent the incandescent light by trying to improve the candle!" The history of sciences pleads for a completely different approach.

Where could this new approach come from?

From our point of view, this new approach should come from a completely different domain of science. Thinking about our products and the development we observe in the science of materials, we realize that the manufacturing process mainly consists in "putting some order" in the material: we have to arrange particles in such a way they fulfill some function which is of interest to us.

Doesn't the concept of "order" remind you something? Probably yes, if you heard about entropy. Entropy is really a measure of "disorder".

## 15.1 Measuring Entropy

A measure of entropy was proposed by Ludwig Boltzman in 1896:

$$S = k \ln \aleph$$

where $k$ is the Boltzman constant, $k = 1.381 \times 10^{-23}$ J/K and $\aleph$ (see Ref. [5], p. 84) is the number of different ways in which the energy of the system can be achieved by rearranging the atoms or molecules among their available states.

An example is mentioned by Atkins [5]: what is the entropy of a crystal containing 1 mol ($6.022 \times 10^{23}$ molecules) in which a molecule can adopt two orientations. Then $\aleph = 2^{6.022 \times 10^{23}}$ ($6.022 \times 10^{23}$ is the Avogadro number) which gives an entropy of 5.76 J/K. Although the number of states is extremely high, entropy is a quantity easy to handle.

## 15.2 Application to Manufacturing

Starting from a block of raw material which has a large amount of disorder, and therefore a high entropy, manufacturing consists in decreasing this entropy: changing the dimensions, managing tolerances, finishing the surface, etc.

Reducing the entropy of a piece of metal can be done by adding some work to it. And the work performed is precisely the cost we are interested in.

Using the entropy naturally solves the question of the product size, so often met in specific and general models, because entropy depends on the product size, via $\aleph$.

The problem would be to measure for old products, or forecast for new products, the amount of entropy, remembering the higher the order, the smaller the entropy.

The nice thing about entropy is it does not care if the product is mechanical or electronic. One can add that, as it goes down to the atom level we can still use it when our technology improves, which means when we are able to arrange particles at a smaller and smaller degree.

## 15.3 About Materials

The concept of entropy can easily be applied to materials: when reading articles about the improvements made on materials, one realizes that these improvements always come by adding some order in raw materials. Examples are numerous, from transistors, integrated circuits to nano-technologies.

## 15.4 About Construction

A new house has less entropy than the materials it was built from. A hurricane shows what happens to a house when its entropy does increase.

## 15.5 About the Maintenance

The second law of thermodynamics says the entropy of an isolated system increases in the course of a spontaneous change:

$$\Delta S > 0$$

which means that the amount of disorder increases with time.

Isn't it exactly what we see when we used equipments? They worn out during the time, which means that their entropy increases. In order to be able to use them again, we have to reduce this entropy, which means re-adding some work in it: this is the concept of maintenance!

Consequently entropy is a very interesting unifying concept in the concept of cost estimating.

# Part IV
## Using Models: Parametric Models Within a Business

## Part Contents

Chapter 16 **Using a Specific Model**

Chapter 17 **Using a General Model**

Chapter 18 **Introduction to "Risk" Analysis**

Chapter 19 **Auditing Parametrics**

Cost models do not exist in a vacuum. They are developed and/or used inside a business for making estimates. Inside this business they are just a tool; this tool offers no guarantee that the cost estimates will be reliable. It is the set "tool + process + environment" which, together, contribute to the cost estimating efficiency.

This introduction gives some recommendations in this respect.

**Building the Environment**
The environment is the set of:

- an organization,
- a personnel selection and the training of these persons,
- a personnel management,
- an audit track.

We limit our comments here to the decision phase.

There is no need to emphasize the role of the organization. Cost estimating is, in this respect, a common function. However, there is a decision to be made about the organizational structure. As soon as a business becomes a bit large, it is organized in departments, projects, etc. each one having some autonomy. This implies to make a choice for the place of the cost estimators. These people can be:

- *Either centralized*: an office is created in which all the cost estimators work together for the departments or the projects.
- *Or decentralized*: the cost estimators are spread in the departments and projects which need them.

Both solutions have advantages and disadvantages. The choice must not forget two facts:

1. Cost estimating is now (especially during the decision phase) a technique (equal to all the techniques used by engineers) which requires more expertise than experience (it may be different during the validation phase: experience is then probably more important). Consequently both solutions are neutral in this respect.
2. Cost estimators know nothing about the projects they have to estimate: they must get the information from other people.

**Centralization** has obvious advantages (the work load can be easily distributed, people may help each other, …). Other important advantages are as follows:

- There will be one data base only.
- It is easy to control how the cost estimates are made: one can expect that all cost estimates are made about the same way. This improves a lot the credibility of the estimates inside the company.
- Training is easier to organize, as well as vacations and other absences.

The major disadvantage of centralization is psychological: cost is not a "neutral" information. It is really a measure of individuals' efficiency. And the cost estimating office – because the cost estimators will have to compare estimates and actual costs – may be perceived as 1. irresponsible (because they will not have to accomplish the job they estimate) 2. critics of the line managers. The consequence is that these line managers will either give no information, or give poor information.

**Decentralization** has the corresponding advantages. The major disadvantages are:

- There will be several databases, as each cost estimator will build the one he/she strictly needs.
- Credibility is low because nobody knows exactly how the estimates are really prepared.
- Cost estimators tend to have their own methods.

The choice between centralization and decentralization must therefore be done with care, and the disadvantages must be palliated. For instance in the hypothesis of decentralization, cost estimators must met frequently in order to compare their methods and make sure they work about the same way. If centralization is adopted, the line managers should be consulted.

A mix solution can also be selected: in this mix solution, cost estimators are centralized, but they are "detached" to the projects which need them for the period of the estimate. Consequently they will work close to the line managers.

One of the major problem with personnel management is related to the careers businesses can offer to cost estimators: How can a company attract good people for this important function? It does not seem that this problem is solved.

## Building the Process

A tool, by itself, does not make estimates. As emphasized by the Software Engineering Institute (SEI), the cost estimating work must be considered as a process. This process involves six steps:

1. To get a reliable database about the projects already made by the company.
2. To use a reliable method for quantifying the characteristics of the object to be estimated.
3. To use a known method for extrapolating from past results.
4. To control these methods.
5. To take into account the constraints of the projects.
6. To organize the feedback of information from the projects.

## About the Tool

By "tool" we mean either a tool for building specific models (this tool must include data analysis, metric selection, type of formula, etc.) or a general model (ready to be used).

This tool, which responds only to step three, can be made inside the company, or bought.

The advantages of buying an existing tool is to benefit from other people experience. This solution may appear expensive, but, from our experience, it will cost much more to develop internally this kind of tool (if the personnel is valorized at its right cost).

We return now to the technical sides of the cost estimating.

Using a specific model is completely different from using a general model: both utilizations must be distinguished and are the objects of Chapter 16 on one hand, Chapter 17 on another hand. These parts deal with cost estimating of the "nominal" project, which means the project as it can be represented by a nominal description of:

- what is to be realized,
- how it will be done.

After that we discuss what is generally called the "risk". No project is really nominal. This does not mean that the nominal project is purely ideal: it means that some realistic hypotheses were made and that it is assumed that these hypotheses are correct. Cost estimates cannot be perfect. There are three causes to this lack of perfection:

1. Partial lack of exactness due to the cost estimating methods. This is "built-in" in the method we use; it may come:
   – from the data themselves,
   – from the methods used to extrapolate from these data.
2. Imprecision related to the nominal project description (the value of the parameters).
3. Uncertainty about what is going to happen.

When we make an estimate, we must always think at the person who is going to receive it. This person is interested of course by the cost we estimate, but beyond that he/she must know how much this value is reliable on one hand, what is going to happen if the project is not nominal on the second hand, and eventually he/she may want to make some trade-offs analysis between performance and cost.

What we have to do as cost estimators is to give, beyond the nominal cost, to the person who receives the estimate the information he/she really needs.

We finish this part by some comments about auditing parametrics.

# 16 Using a Specific Model

## Summary

This chapter deals with the usage of a specific model for estimating the cost of a product belonging to the product family (its population) for which the specific model was built.

It assumes that the values of all the causal variables are known with precision and certainty: taking into account imprecision and uncertainty will be dealt with in Chapter 18.

Even if the values of the causal variables are perfectly known, the cost cannot be, due to the fact that its value will be computed with a formula of which coefficients are only partially known with a perfect accuracy. It can be seen, in Volume 2, that their standard error can be computed.

This is the first cause of cost inaccuracy, which is fundamental because it cannot be escaped. This chapter will establish how this inaccuracy can be computed.

The study of the other causes of inaccuracy is postponed up to Chapter 18.

Specific models have been built. The problem is now to use them in order to estimate the cost of a new product. This implies that a specific model is selected.

### The First and Important Questions

Does the product really belong to the product family for which the model was built? The question seems obvious but experience shows it has to be asked: the fact that the product to be estimated bears the same name does not, by itself, proves this belongingness.

The second question is: Does the estimator know the values of the parameters which are included in the model? Is he/she sure that the available parameters have the exact meaning chosen by the cost analyst when the model was built? These parameters are called (the index 0 is generally used in all the manuals): $X_{1,0}, X_{2,0}, \ldots, X_{j,0}, \ldots, X_{J,0}$.

These questions are not difficult to answer if the model is properly documented by the cost analyst.

*What Is the Result of a Cost Estimate?*
The result of a cost estimate should be a set of three things:

1. A nominal cost, called $\hat{Y}_0$, which is purely theoretic and should therefore never published alone.
2. The precision with which this nominal cost is known. This precision is generally given by a confidence interval in which we can expect the true cost to be 90% or 95% of the time (this is called the level of confidence).
3. The sensitivity of the nominal cost to a small change of each causal variable, which is given by the partial derivatives $\partial \hat{Y}_0 / \partial \hat{X}_j$.

The first two things, both being equally important, tell "what we can say about the cost of this future product, given the data we have for similar products (our product family), given the values of the causal variables".

The third thing is also important: it tells the user of the cost estimate what can be expected if a small change (generally 5% or 10% is selected) of any cost driver is considered. This user will assess the credibility of the estimate on this sensitivity, which is also the input of any trade-off analysis.

Therefore to any nominal cost estimate should be joined the sensitivity of all the parameters. This sensitivity will allow the person who receives the estimate to make a preliminary judgment about the risk he/she takes if he/she accepts the estimate as it is: if he/she sees that the cost is very sensitive to one parameter, he/she must immediately question about how was determined the value of this parameter from which the cost is estimated.

He/she will also immediately consider trade-offs, especially if the parameters are mainly functional in nature, and therefore express the demands made by the customer on the product: how is the cost going to change if is it possible to reduce such or such a demand?

The values of the variables $X_{1,0}, X_{2,0}, \ldots, X_{j,0}, \ldots, X_{J,0}$ must of course be also given, as well as the precision with which they are known (estimated!). The way this precision can be taken into account is described in Chapter 18.

## 16.1 The Classical Approach

The classical solution is based on the Gauss' hypotheses.

*Notations*
In order to avoid any confusion, we call the values computed from the sample $\bar{\hat{B}}_0$, $\bar{\hat{B}}_1$, etc. keeping the notation $\hat{B}_0, \hat{B}_1, \ldots$ for the true estimators of $B_0, B_1, \ldots$. $\hat{B}_0, \hat{B}_1, \ldots$ are random variables, of which $\bar{\hat{B}}_0, \bar{\hat{B}}_1, \ldots$ are the mean, their variances having been computed by the formulae given in Chapter 15 of Volume 2.

The purpose of this chapter is to estimate the cost of a new product belonging of course to the same product family. The values of the causal variables are given by the set[1] $X_{1,0}, X_{2,0}, \ldots, X_{j,0}, \ldots, X_{J,0}$.

---

[1] Capital letters are used for objects of the population (which means not included in the sample).

# Using a Specific Model

The estimated cost is called $\hat{Y}_{0,est}$:

$$\hat{Y}_{0,est} = f(\hat{B}_0, \hat{B}_1, \ldots; X_{1,0}, X_{2,0}, \ldots)$$

## 16.1.1 Using One Parameter Only

This section deals with one parameter only. The purpose of describing this usage is that it does not require the use of matrices and therefore the reader is able to see all the computational details.

### The "Nominal" or "Mean" Cost

The nominal cost is simply given by the value of the dynamic center for the value of the parameter. This is true whatever the type of formula.
This value is generally called $\hat{Y}_0$:

$$\hat{Y}_0 = f(\bar{\hat{B}}_0, \bar{\hat{B}}_1, \ldots; X_{1,0}, X_{2,0}, \ldots)$$

### A First Level Cause of Imprecision

As we already mentioned it several times, a specific model is the set of the dynamic center (which gives the nominal cost), plus the distribution of the deviations around this center. These deviations produce the fact that the estimated cost of the new product is only known with some imprecision, imprecision we want to estimate as well. If we could manufacture several products with the same set of parameters $X_{1,0}$, $X_{2,0}, \ldots, X_{j,0}, \ldots, X_{J,0}$, the nominal cost will only be the average of their costs.

Due to these deviations the estimated cost is the sum of the nominal cost, plus some imprecision related to these deviations, imprecision around this nominal cost.

Of course we do not know the deviations for all the individual products of the population. We know only the residuals (which are not the same things) in the sample, residuals computed on an estimated dynamic center. As it can be seen in Chapter 15 of Volume 2, these residuals are used to estimate the variance of the deviations.

Let us start with the bilinear case with only one causal variable, simply called here $X$:

$$\hat{Y}_0 = \bar{\hat{B}}_0 + \bar{\hat{B}}_1 X_0$$

where $\bar{\hat{B}}_0$ and $\bar{\hat{B}}_1$ are here the "nominal" values of the intercept and the slope, $X_0$ the value of the parameter for the new product. As we know it, these nominal values are the values computed from the sample ($b_0$ and $b_1$).

In this formula $X_0$ is a well defined number, whereas $\hat{B}_0$ and $\hat{B}_1$ are in fact random variables, as they would vary from one sample to another one (each sample is supposed to give a different set of values $y_1, y_2, \ldots, y_i, \ldots, y_I$) if we were able to draw several samples, keeping the $x_i$ constant. Another way of saying the same thing is that $\hat{B}_0$ and $\hat{B}_1$ are given by:

$$\hat{B}_0 = \frac{1}{I}\sum_i y_i - \hat{B}_1 \times \frac{1}{I}\sum_i x_i = \bar{y} - \hat{B}_1 \bar{x}$$

$$\hat{B}_1 = \frac{\sum_i (x_i - \bar{x})(y_i - \bar{y})}{\sum_i (x_i - \bar{x})^2}$$

and their variances and covariance by:

$$\text{var}(\hat{B}_1) = \frac{S^2_{E+}}{\sum_i (x_i - \bar{x})^2}$$

$$\text{var}(\hat{B}_0) = \frac{\sum_i x_i^2}{I\sum_i (x_i - \bar{x})^2} S^2_{E+}$$

$$\text{cov}(\hat{B}_0, \hat{B}_1) = \frac{-\bar{x}}{\sum_i (x_i - \bar{x})^2} S^2_{E+} = -\bar{x} \times \text{var}(\hat{B}_1)$$

where all the $x_i$ are fix values, whereas the $y_i$ are random values. This is part of the Gauss' hypotheses: we assume that we could be able to draw different $y_i$ for the same $x_i$.

The distributions of $\hat{B}_0$ and $\hat{B}_1$ follow a normal distribution in which the standard deviation $S_{E+}$ of the deviations in the population appears. If we replace this unknown value by an estimate, we saw that the coefficients follow now a Student distribution with $I - 2$ degrees of freedom ($I - 1$ if the intercept is forced to be 0) always around the true values $B_0$ and $B_1$.

Therefore the true value of $Y_{0,est}$ must be computed with the same formula as $\hat{Y}_0$, but now with the true values of $\hat{B}_0$ and $\hat{B}_1$.

Consequently $Y_{0,est}$, being the sum of random variables, is also a random variable: its value is going to change from one sample to another one. As for any random variable, we can define its distribution by its mean and its standard deviation.

### The Distribution of $Y_{0,est}$

The **mean** of the distribution of $Y_{0,est}$ is given by $\hat{Y}_0$.

What about its **standard deviation**? The computation is more difficult, but easy to follow if we compute the distribution of the estimate error. The "true" value of the cost of the new product is given by:

$$Y_0 = B_0 + B_1 X_0 + E_+$$

where $B_0$, $B_1$ and $E_+$ are the true (but unknown) values. This gives a prediction error equal to:

$$Y_{0,est} - Y_0 = (\hat{B}_0 - B_0) + (\hat{B}_1 - B_1)X_0 + E_+$$

of which variance is given by:

$$\text{var}(Y_{0,est} - Y_0) = \text{var}(\hat{B}_0) + X_0^2 \times \text{var}(\hat{B}_1) + 2X_0 \times \text{covar}(\hat{B}_0, \hat{B}_1) + S_{E+}^2$$

and therefore:

$$\text{var}(Y_{0,est} - Y_0) = S_{E+}^2 \times \left[1 + \frac{1}{I} + \frac{(X_0 - \bar{x})^2}{\sum(x_i - \bar{x})}\right]$$

and we therefore conclude that the variance of $Y_{0,est}$ around $Y_0$ is given by:

$$\text{var}(Y_{0,est}) = S_{E+}^2 \times \left[1 + \frac{1}{I} + \frac{(X_0 - \bar{x})^2}{\sum_i (x_i - \bar{x})^2}\right]$$

According to hypothesis ④ (see Chapter 15 of Volume 2), the distribution of the deviations is normal. $Y_{0,est}$ being the sum of normally distributed values also follows a normal distribution. Consequently its **skewness** is equal to 0 and its **kurtosis** to 3.

The variance of $Y_{0,est}$ include the term $S_{E+}^2$ which refers to the population and which therefore we do not know. As we did previously we want to replace it by its estimate knowing that if $A$ represents a random variable following a normal law $N(0, 1)$ and $B$ a random variable following a $\chi_q^2$ distribution (described in Chapter 3 of Volume 2), then the ratio $A/\sqrt{B/q}$ follows a Student distribution with $q$ degrees of freedom.

It so happens (from what has been said in the second paragraph of this section) here that $Y_{0,est}/\text{var}^{0.5}(Y_{0,est})$ follows a $N(0, 1)$ distribution and that $((I - 2) \times s^2)/S_{E+}^2$ follows a $\chi_{I-2}^2$ distribution; therefore:

$$\frac{Y_{0,est}}{S_{E+} \times \left[1 + \frac{1}{I} + \frac{(X_0 - \bar{x})^2}{\sum_i (x_i - \bar{x})^2}\right]^{0.5}} \times \sqrt{\frac{S_{E+}^2}{(I-2) \times s^2} \times (I-2)} = \frac{Y_{0,est}}{s \times \left[1 + \frac{1}{I} + \frac{(X_0 - \bar{x})^2}{\sum_i (x_i - \bar{x})^2}\right]^{0.5}}$$

follows a Student distribution with $I - 2$ degrees of freedom.

The variance of this distribution can therefore be computed from this distribution and therefore the variance of $Y_{0,est}$, when one uses $s^2$ instead of $S_{E+}^2$ is given by:

$$Y_{0,est} = Y_0 \pm st(I - 2, 1 - \alpha/2) \times \left[1 + \frac{1}{I} + \frac{(X_0 - \bar{x})^2}{\sum_i (x_i - \bar{x})^2}\right]^{0.5} \times s$$

where $\alpha$ represents the level of confidence (for instance 90%) and $st$ the student's distribution.

*About the Distribution of an Individual Estimate*

What can be said about this distribution? One first observes that the standard error of the estimate depends on $X_0 - \bar{x}$ which is the distance of the causal variable value of the new product from the average value of the causal variables in the sample: the closer is the new product from this average value, the smaller is the standard deviation. This quite understandable from a cost estimator point of view: in the vicinity of the average value, all data points in the sample contribute symmetrically to the estimate; on the opposite, when we estimate a new product far away from this average, not so many data points contribute to it.

The distribution depends also on $I$, the number of data points in the sample. This is also quite understandable and the formula just quantifies our intuition.

And (this was a bit unexpected) it depends on the sum $\sum_i (x_i - \bar{x})^2$ which is closely related to the variance of the $x_i$ in the sample: the larger this variance (how much the causal variable is scattered), the smaller the standard deviation of the estimate. Consequently we must try to get the $x_i$ as spread as possible. When one thinks about it, there is also some logic in this statement: it is better for the quality of the estimate to have the $x_i$ spread instead of having them concentrated. If it were possible, we should therefore select data points for which this variance is maximum: but this is a rather theoretical approach when dealing with cost!

*Example*

Figure 16.1 displays both the "nominal" cost of any new product belonging to a given product family, as well as the confidence interval (with a level of confidence of 90%) of the cost estimate. This confidence interval is represented by the two dotted lines (which are hyperbolae).

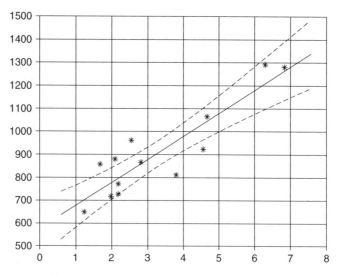

**Figure 16.1** Example of the distribution of $Y_{0,\text{est}}$.

The relationship is given by:
$$\text{Cost} = 573 + 101 \times \text{Mass}$$

One realizes on the figure that, for this number of data points and their spread around the dynamic center, the confidence interval doubles from the middle of the data to the extreme values.

### 16.1.2  Using Several Parameters

The logic is exactly the same, it is also based on the Gauss' hypothesis. $\vec{X}_0$, the set of the causal variables for the new product, can now be described as a vector. The variance of $Y_{0,\text{est}}$ is now given by:

$$\text{var}(Y_{0,\text{est}}) = S_{E+}^2 \times X_0^t \otimes \left( ||^+x||^t \otimes ||^+x|| \right)^{-1} \otimes X_0$$

and the discussion about the standard error of the estimate is the same.

### 16.1.3  What Can be Done When the Number of Data Points Is Limited?[2]

When the number of data points is limited, the number of parameters is necessarily limited. We already mentioned that a rule of thumb, in order to have a realistic figure for the deviations from the dynamic center, is that about five data points should be made available for each parameter.

In the situation mentioned here, no more than one or maybe two parameters can be used. This means that, if the product family is not strictly homogeneous, the deviations may be important and we have no way to reduce them; these deviations are certainly not "errors" as they are only due to the impossibility to use more parameters, even if they exist.

This is not rare in the cost estimating environment: the data points which are available for building the model are sometimes not very numerous and the product family cannot be considered as fully homogeneous. A solution has to be found, especially if we are confident in the costs which are in the data base.

The formula and the distribution of the deviations which are computed cannot really be called a cost estimating relationship (CER): a better name would be a cost estimating trend (CET) because the number of data points is too small to get a great confidence in the formula.

But the cost estimator has to work with it! He/she can then proceed in four steps:

1. The nominal cost is computed, as usual from the dynamic center. Suppose this cost is €50K.
2. Then he/she has a look in the database in order to find out a product which is rather similar (in terms of technology) to the product which has to be estimated. Suppose he/she is confident in the cost (let us say €150K) which is in the database for this similar product.

---

[2] This approach was suggested to us by Alain Buck and his staff (Alstom Marine, Saint Nazaire, France).

3. Having a look at the residuals, he/she discovers that this cost is 12% higher than the dynamic trend.
4. Due to the fact this cost is considered realistic, he/she then decides to apply the same percentage as a correction to the value given by the trend. The estimate then becomes €56K. It is, if we are confident about this similar product cost, more realistic than the first estimate.

Another approach, which we used from time to time, is a step by step approach. Suppose we have 6 data points and 4 parameters we considered as interesting. We start by building a model with the most interesting – from a cost estimating point of view – parameters. From this model we can compute residuals; now comes the second step: we try to "explain these residuals with the two remaining parameters (of course the process could be repeated). This solution is not perfect from a strict mathematical point of view, but it may help. Quite obviously the standard error of the estimate is a little bit more complex to compute, but it is feasible.

## 16.2 The Modern Approach

The modern approach is based on the Jackknife or the Bootstrap, which are described in Volume 2.

For cost estimating purposes, unless we get a lot of data, the Bootstrap is generally more convenient. As it was said, it consists in generating a lot (let us say 400) of replications of our sample, based on a random selection, repetitions being allowed, of the data in the original sample.

From each of these replications a formula may be computed, from which the model can be derived. In this model, due to the fact that no hypothesis is made on the data, the distribution of the $\hat{B}_0, \hat{B}_1, \ldots$ may be not normal, which means that its skewness is not equal to 0 and its kurtosis not equal to 3.

For cost estimating a new product, two solutions are possible as follows:

1. Either starting from the model and the distribution of the $\hat{B}_0, \hat{B}_1, \ldots$ it is possible to derive an estimate and its standard error.
2. Or starting from the 400 values of the $b_0, b_1, \ldots$ one can compute 400 estimates; therefore the distribution of these estimates is known: from it the average value, the standard error and even the skewness and the kurtosis of the estimate can be computed.

This second approach should be preferred because the computations are simpler.

# 17 Using a General Model

## Summary

Using a general model is not difficult but requires a good understanding of the model. This chapter, remaining at a high level of generality, gives some information about the way a model can be used.

As the book does not refer to any model in practice, the comments have to be rather general.

The chapter starts with a preliminary comment about the confidence interval which may be expected from the model. This is true for any model.

The major conclusion at this stage is that, whatever the type of model you intend to use, there is no way to escape from some definition of the product to be estimated. Just knowing its name is certainly not a product description.

## A General Comment

Using a general model is rather basic. There are of course some comments about all these models. The most important comment is that these models do not provide the level of confidence of the estimate, because the model builder cannot know how the model will be used. This means that this level of confidence must be established by the model user. The only way to do it is to use existing products and to compare what the model "says" with the known cost. Two outputs are the consequence of this comparison:

- First the user learns what it means to be consistent when using a model.
- Second he/she learns what standard error of the estimate can be expected. When the model disposes a variable which has to be adjusted when making an estimate, and which is computed when the cost is given, the standard error of the values found from existing costs immediately gives the answer. This is an advantage; otherwise the user has to work directly from the comparison between estimated costs and observed costs.

This also means that the quality of the estimate depends a lot on the model user. It must be reminded here that this person must describe the product to be estimated (this is not the case with the specific models because, for these models, a product is defined by its belongingness to a product family, plus the quantification of the differences with the "reference" product). The cost estimator gets information about the product from other people, who use their own language. As the model also has its own language, the work of the cost estimator is really to *translate information from one language to another one*.

The more general the model, the more difficult is this translation. For this reason, using a type 3 model is more difficult than using a type 2 model (this is the price to pay for generality), and using a type 2 model is more difficult than using a type 1 model.

But the difficulty must not be exaggerated. Once the model is properly understood, it only requires a reliable, which does not mean detailed, description of the product on one hand, consistency in the approach one the other hand. Consistency is really a key word when using a general model.

## 17.1 Using a Type 1 Model

Using a type 1 model does not seem to present any difficulty: the choice of the specific model to be used is relatively simple, if the definition of the models is well documented. Problems may appear if the definition of the product to be estimated either is not clear enough or does not fit exactly with any specific model.

We nevertheless saw that the builder of a model of type 1 has to add a qualitative variable, in order to overcome the difficulty of using a limited set of models. This qualitative variable offers a choice; we often discover that this choice change the nature of parametric cost estimating. This comes from the fact that any experienced cost estimator has a preconceived idea of what the cost should be – not exactly of course, but enough to influence his/her judgment about the result he/she gets. So the estimator starts by selecting one modality of the qualitative variable and looks at the result. If he/she finds that the result does not fit at least approximately with his/her preconceived idea, he/she adjusts the qualitative variable until an "acceptable" result is obtained; it is not difficult then to "rationalize" and get a satisfactory explanation for this adjustment. Then the result is published with the comment: "the model says so …".

This is, of course, not the way a model should be used; as a matter of fact if the model is just there for finding out the value of the parameters which gives the "correct" answer, its interest is very low.

As this question is not specific to the model of type 1 (but is a normal trend when having to adjust a qualitative variable), we will return to this question in the Section 3.

## 17.2 Using a Type 2 Model

Let us remind the reader that a general model of type 2 is an extension of the Chilton's law and that, in order to be as general as possible, it has to use the mass for quantifying the product size and that the exponent of this mass is supposed to remain constant inside a product family.

Using the mass is a constraint for the following reason: it does not allow to estimate the cost of something which has no mass (such as an activity), or for which the mass is irrelevant (such as a plant), or is not known. This constraint reduces the generality of such a model, even if it must be recognized that most of the products (especially for products which have to be moved) have a mass. Nevertheless using the mass for quantifying the product size is not common in many industries, and having an idea of the mass in the early stages of a project is then difficult.[1]

Then, and unless some correlation might be established with other parameters, the exponent of the Chilton's law has to be found. This requires to know at least one product, and its cost, belonging to the same product family as the product to be estimated.

The first question is then: Are we sure that both products belong to the same product family? If we do not have a (as good as possible) description of both products, it is very difficult to answer the question. This is particularly true in the early phase of a project.

The second question is, when some differences are known between the products, that the exponent has to be adjusted and this is very difficult to do, unless you have established the mentioned correlation. As this correlation must be based on the product description, we also need it here.

## 17.3   Using a Type 3 Model

The user must first decides, if the model allows it, about the options which are to be used. We believe that the reality is too complex and diverse to fit with just one type of model: a general model should be considered as a set of several models among which the user chooses as a function of the problem he/she has to solve. For instance we mentioned there are several types of production organizations; this is an important input for estimating this production cost. But there are many other options which can be included in the model.

Then he/she has to describe the products:

- *The product size* is always something important and the user has a choice: should he/she use functional or physical descriptors?
- *The product structure*, which requires to get some knowledge of the technologies which are used internally. Experience shows that this is not too difficult and, by the way, it is the way value engineering works.
- *The quality level*, either from a predefined scale or by computing it from the distance to the elastic limit the product is going to work at.
- *And eventually the culture of the organization*, at least of the industry type, the product will be manufactured by. If it is possible (this is another option) this culture may be computed from a reference product (which does not have to belong to the same product family) manufactured by the same company or at least by the same industry.

This type of model does not eliminate the criticism which was mentioned in Section 1 of this chapter: the user may "adjust" the model (its parameters) in order to get the cost he/she expects. In order to avoid this behavior, strict rules must be observed:

- The user should first have a good knowledge of the model and as much experience as possible.
- Parameters should be quantified without making any computations with the model. This step may take some time: the user, after collecting information from

---

[1] This is why we insisted on using functional parameters in these early stages.

the design engineers, should make a preliminary quantification of the parameters. Then he/she returns to the design engineers and tells them: "this is what I understood from what you said"; he/she listens to the observations they made and makes a definite quantification (at the same time he/she should collect information about the precision the parameters are defined).
- Eventually the model is run and the results obtained. The user must not at this time make any change, even if he/she does not like the results, unless a typing mistake was made of course.

Now one can really say: "the model, used the following way, says so ...".

This "strategy", which requires a good discipline, is the only one which gets full benefit from the model.

# 18 Introduction to "Risk" Analysis

## Summary

The previous chapters and Volume 2 deal with cost estimating of the "nominal" project which means the project as it can be represented:

- by a nominal description of what is to be realized,
- by a nominal description of how it will be done,
- assuming that everything is going to be all right.

We already saw that this cost has a certain level of inaccuracy, due to the fact it is computed using a sample, which bears on a limited set of values, and these values may be more or less scattered.

But no project is really nominal. This does not mean that the nominal project is purely ideal: it means that some realistic hypotheses were made and that it is assumed that these hypotheses are correct.

There are a lot of reasons for a project to be more or less different from what is expected. The consequences of the differences are generally called "risk".

Cost estimates cannot be perfect. There are three reasons for this imperfection:

1. Inaccuracy due to the cost estimating tool which is used for preparing the estimate. This has already been dealt with when discussing estimating methods; it may come:
   - from the data themselves (how scattered they are),
   - from the methods used to extrapolate from the data.
2. Imprecision related to the parameters describing the nominal project.
3. Uncertainty about what is going to happen during the project life.

Dealing with risk can be, and is sometimes, done manually. This is quite possible for small projects but would be difficult for large projects.

As mathematical tools for carrying out this analysis are now well known and easy to implement even on desk computers, we will develop some mathematical concepts about it. The purpose is not to improve the "accuracy" of risk analysis (this word has little meaning here), but to carry it out very quickly letting the user think about the inputs and the outputs, forgetting the translation from the first ones to the second ones.

Both imprecision and uncertainty are dealt with in this chapter.

## 18.1 Introduction

"Risk" has become a very popular subject: everyone making a cost estimate feels obliged to add a risk analysis at the end of his/her cost estimate. This is fine if the "risk" is properly defined and handled.

This section therefore starts by an "acceptable" (from a cost estimator point of view) definition of risk and briefly looks at the way it is generally handled.

### 18.1.1 A Definition[1]

The term is familiar, but, when you think about it, seems difficult to define with some accuracy. Let us examine the few definitions which can be found in the literature.

The handbook of the Defense Systems Management College (US Department of Defense), "*Risk Assessment Techniques*" quotes a memorandum of former Deputy Secretary of Defense Frank C. Carlucci III, in 1981 (Ref. [19], p. I-1): "the memorandum requires the services (among other things) to incorporate the use of budget funds for technological risk and recommended an increase in Department of Defense (DOD) efforts to quantify and expand the use of budgeted funds to deal with uncertainty". The memorandum uses the words "risk" on one hand, "uncertainty" on the other hand; it does not imply they are equivalent.

However, the handbook (Ref. [19], p. II-2) defines **risk** as "the term used to denote the *probability of an event* and its consequence"; more specifically, it says that "the concept of risk in this handbook is that of the *probability* and consequence of not achieving some defined program goal (such as cost, schedule or technical performance)". In other words, this handbook considers that "risk" and "probabilities" are strongly related, if not synonymous, terms, or that the only way to deal with risk is the language of probabilities.

We agree with the handbook on one point: risk has to do with the **uncertainty about achieving some defined program goal** (we will here limit the discussion to one major goal of any project: its cost). Without uncertainty of any kind, there would be no risk.

But, as it is said in the introduction, we disagree with the common assimilation of risk to probabilities. In order to clarify the subject, let us have a review of the definitions used by several other authors:

- Larry D. Gahagan (Ref. [28]) says that "cost risk is concerned with three things: the randomness of the unplanned and unknown events potentially facing a system development program, their **probability of occurrence**, and the consequences of their occurrences".
- Donald W. MacKenzie (Ref. [37]) adopts the definition given in the *AFSC Cost Estimating Handbook*: risk is the outcome subject to an uncontrollable event stemming from a *known* (underlying) **probability distribution**; uncertainty is the outcome subject to an uncontrollable event stemming from an *unknown* (underlying) **probability distribution**. He also recognizes that *the most important source of uncertainty in a cost computation is associated with uncertainties in the individual input parameters*.

---

[1] This paragraph was presented by the author in the ISPA conference of 1996 (Boston).

Introduction to "Risk" Analysis 255

- For John Bowers (Ref. [12]), risk is that fatal combination of uncertainty and impact; about uncertainty, he adds: "the uncertainties can be described using **a variety of probability distributions**".
- Gerald R. McNichols, in an article entitled *"An historical perspective on risk analysis"* (Ref. [42]), writes that his article presents *"Historical generation of parameter values in **subjective probability density functions**"*, but adds "this is only one small part of the total problem of treating uncertainty in parametric costing"; unfortunately, he does not comment on this total problem.

It is quite obvious that, for all these authors, risk and probabilities are two faces of a same reality and, consequently, that it is impossible to speak about risk without using the theory of probabilities.

Let us challenge this point of view. In order to do so, we need a definition of risk which is not immediately associated with probabilities. We propose then the following definition: "risk is anything that may change the single best estimate that is computed by the program manager". S.A. Book proposed (Ref. [9]) a similar definition in an article published in 1994: *"Risk is the assessment of inadequacy of current and projected funding profiles to assure that program can be completed and meet its stated objectives"*.

This definition immediately rises a problem: it is well known that any cost estimate is based on comparisons between the object to be estimated and our past experiences. The latter includes obviously programs in which the amount of risk was not negligible; therefore any "single best estimate" includes a part of risk,[2] unless the historical costs are normalized to non-risk situations before being used to build a cost estimating relationship (CER) or to adjust a general model. As this normalization process is extremely rare if non-existent, one could say that adding a new risk over the estimate may count the same thing twice. This is not impossible and must be investigated when using historical data.

Before we detail the "risk factors", let us observe that some "risk drivers" people generally consider, are not risk drivers at all and should then be already included in the "single best estimate". Examples: schedule constraints, budget constraints, cost growth due to technology improvement, toxic materials, etc. All these points can be "modelized": there is no reason not to include them in the estimate. Quite obviously, these factors have to be known; if they are not sure, they must be dealt with as "alternatives".

Let us now investigate the "risk factors". We think it is possible to classify them into three classes for which we will use (the choice of the words is deliberate: instead of creating new words nobody will understand, we prefer to use common words with a specific meaning) the following words:

- "uncertainty" (of the program definition);
- "imprecision" (about some inputs);
- "chance" (related to the program path).

Let us briefly discuss these three classes.

---

[2] We do not agree with some people opinions saying that the risk of past experiences appears in the residuals we can observe after a regression analysis, for two reasons: first the regression analysis develops relationships based on an "average risk" (this is the purpose of the regression analysis); secondly, this mixes uncertainty on the values of the inputs and "chance" (the term is defined later on).

### 18.1.2 The Components of Risk

*Uncertainty*

Every estimator, in front of a new program to be estimated, needs a scenario describing how the project is supposed to be carried out. This scenario, as we mentioned it before, must include the constraints and other circumstances which are part of the project description.

This scenario generally considers the *product description* on one hand, the *development/production environment* on the other hand.

There may be many "unknowns" about the scenario; these unknowns express the project "uncertainty". The scenario implies many (discrete) choices from the engineers and other people; the choices cannot be made at the time the estimate is carried out, but we know there will be some decisions to make in due course, once some information is made available. Let us give some examples of such unknowns:

- Someone may consider using one of a set of alternative technologies, depending on their availability, the need to reduce the mass, the procurement policy, etc.
- The foundation of a building may be more or less complex due to the nature of the ground as will be revealed by the geologic survey.
- A software may be more or less complex according to the hardware people ability to fulfill some of the specifications. At the beginning of a project, trade-off analysis may help decide on what will be achieved by hardware and what functions will be devoted to software; as the project goes on, it may appear that the hardware developers cannot achieve their goal. They then quite often ask for a modification of the software; as they say: there are only a few lines of code to change!
- The experience of the people may differ depending on the contractor who will get the contract.
- And so on.

All these different scenarios must be dealt with as alternatives: it is quite possible to develop as many estimates as there are alternatives in the project. Each alternative may be assigned a (subjective) probability: for example, the probability that this technology will be used is 0.7, the probability we will find rock in this location is 0.4, the probability we will have to increase the software due to hardware difficulties is 0.3, the probability this contractor will get the contract is 0.8, etc.

Techniques to handle such situations are well known (decision trees for instance).

The probabilities mentioned here are all discrete because they are attached to alternatives. They can be easily handled by statements such as "if … then … else", even if there are many alternatives and if some of them depend on other ones. This is a classical problem of probabilities which does not require, except in exceptional circumstances, expensive (in term of processing time) tools such as the Monte-Carlo simulation.

*Imprecision*

Imprecision describes a different situation.

Let us take an example (among many others): many CERs use the mass – or other technical specification which will be translated into input parameters – to quantify

Introduction to "Risk" Analysis 257

the size of an equipment (or the number of statements of a software, or the square footage of a building). In the early stages of a project, it is difficult, if not impossible, to quantify this size with sufficient accuracy, because the specifications are known at the system level and that some work will have to carried out to quantify the size at the equipment level (where the CERs are available), or because the transfer function between specifications and size is not known with accuracy.

**This is the most common "risk"** the estimator has to deal with; the theory of probability has been applied to this subject, mainly because there was no other tool available and probably also because this theory is part of our culture.

Nevertheless, is it realistic?

Let us not forget *the basic assumption of probabilities* (this is quite obvious in the Monte-Carlo technique), even for subjective probabilities: using probabilities means that the variables we are interested in is supposed to randomly fluctuate during the project, until its value is definitely chosen. Of course, in these random fluctuations, some values or range of values are considered more realistic than other ones. This is the reason why estimators try to "guess" this distribution and why the possible distribution patterns are so often studied in risk analysis.

We consider that this basic assumption is wrong when we deal with this type of risk. Maybe two or several values should be considered, depending on the technology being considered or due to other constrains, but we cannot agree that, as long as it is not definitely established, it will randomly fluctuate!

We are dealing here with **a fundamental risk**: We know that the size of the product will always be the same, but we are unable to tell exactly what it will be!

## Probabilities "Lie" to the Decision-Maker

Let us examine how the probability distribution is interpreted by the decision-maker.

Suppose for instance that the mass will be between 50 and 60 kg; this mass is used in a CER to estimate a cost. The probabilist will use a certain – beautiful – distribution pattern, whatever it will be, and will present to the decision-maker a cumulative distribution of cost, showing, for instance, that the probability of the cost being less than €95 000 will be 0.8.

The decision-maker, depending on his aversion or not of risk, may decide that fixing the budget of the project – or the price – at this value will give him/her an acceptable level of confidence of not loosing money on this project. If the project was to be made 100 times, and if for each version the mass is going to randomly change, he/she may be right. But if the project is to be made just once and if the interval between 50 and 60 kg only means a normal – at this stage of the project – imprecision about the mass, he/she is wrong as the mass might very well be 59 kg: the problem is that we do not know it.

Cumulative probabilities necessarily imply the basic assumption of probabilities, which is called the "additivity principle". This principle does not apply to our problem: the chances of having the weight between 50 and 56 kg **are not** the sum of the probability of having the weight between 50 and 53 kg, plus the probability of having the weight between 53 and 56 kg.

It is clear that the theory of probabilities is an inadequate and wrong way of solving this problem. How can we solve this problem?

**Fuzzy numbers** are exactly the tool we need: they are constructed on the same principles as probabilities, *except for the additivity principle which is rejected* (this is explained later on). They adequately reflect the situation we are speaking about: the inability to quantify a variable (which will have one value and is not supposed to randomly change) with the information we have. This does not forbid the user to say that the value of this variable is more "possible" in a certain range than in another. This is allowed by the theory of "possibilities" (the theory from which fuzzy numbers are developed) which is shortly described in another section.

## "Chance"

Whatever the human activity, its cost or the time we need to accomplish it are the result of some "random drawing" of one output among hundreds of possible outputs. This has nothing to do with uncertainty or imprecision: it is inherent to human activity. We call it "chance".

Many things may explain this random drawing: if you do the same activity twice, you will not spend exactly the same time to do it, and this is true for the thousand activities which are part of a project. You may experience some minor technical problems (such as those I have this morning with my word processor), you may be disturbed by telephone calls, one supplier may be late to deliver, someone you need may be ill, etc.

Similarly human activities costs are necessarily random numbers and we have to establish their statistical distribution. Due to the number of independent variables which randomize this process, we may expect, from the central limit theorem, to discover there a normal distribution. The standard deviation is not well known – not so many results have been published in this domain – but one may expect a standard deviation between 1% and 5%, depending on the number of people to coordinate to accomplish an activity, the stability of the techniques, etc. including the fact you may be sometimes lucky and sometimes unlucky. This partly explains the residuals in any statistical study of cost experiences and sets a limit on the accuracy of any cost forecast.

There is nothing we can do about it, except taking this fact into account in our cost estimates.

### One Remark About "Chance"

Suppose the project includes 100 activities and to make it simple that these activities are independent, that their cost is €1000 and they are all estimated with a standard deviation of 5%, or ±€50.

As these activities are independent, are subject to a lot of small perturbations, we can reasonably assume that their cost is normally distributed. The project cost:

- will have a mean value of €100K,
- and standard deviation of $\sqrt{100 \times 50^2} = $ €500, due to the additivity of the variances of independent variables.

Consequently the influence of "chance" on the project is, in absolute term, larger than the influence on the activities, but, in relative terms, much lower: the project cost is known with a standard deviation of 0.5% which is really negligible.

This helps explain why the effect of chance on the project cost is generally speaking not considered.

## Conclusion

We think that most authors foster the use of probabilities as the unique tool for quantifying risk. When we read the available literature, it appears clearly that risk assessment techniques are an application of the theory of probabilities. Due to the fact that the relationships between inputs (the information about the program) and outputs (the program goal) are, generally speaking, too complex to be modelized analytically, the standard tool of risk quantification has become Monte-Carlo simulation.

We believe that this theory, and its "standard tool", are only a partial solution to the risk assessment problem and that they are, in most cases, **an inadequate solution**.

The most important technique, as we see it, is the use of **fuzzy numbers**; this technique of fuzzy numbers has been developed for dealing with the imprecision about the inputs, which is the major "driver" of risk. Fortunately enough, the use of these fuzzy numbers is relatively easy and thousand times quicker that the Monte-Carlo technique (with the sample sizes generally used).

Probabilities are an inadequate tool to take into account the imprecision which is built in our cost predictions. More than that: it is **a dangerous tool** to use, because it gives the decision-maker some false confidence about the range of the expected costs through the cumulative distribution. In a certain way, it lies to the decision-maker.

Probabilities have been used in the past because no other tool was available (and probably because our culture makes us feel confident about a result when a lot of nice computations can be made about a subject), but it is now, most of the time, **an obsolete tool in the domain of cost predictions**. The use of fuzzy numbers implies some cultural change, but this change has to be made in order to give the decision-maker the right predictions he/she expects from us. As fuzzy numbers are easy to use, there is no reason not to make this change.

Some cost estimators do not agree with the concept of fuzzy number because, they say, managers, expect from us a probability of cost overruns. They are certainly partially right, because managers have been trained this way. Nevertheless we would like to comment briefly about this idea: managers have to decide in a situation which is partly uncertain and they have to live with this uncertainty. Giving them a probability figure may make them more confident about their decisions, if this figure correctly describes the situation. It is upto you to decide …

## 18.2 Taking into Account the Imprecision of the Model Inputs

### 18.2.1 What Do We Mean by "Imprecision"?

As indicated in the previous section, imprecision of the inputs is the most frequent situation for the cost estimator. The cost analyst prepared a model (specific or general) relating for instance the cost and the product mass, plus any other variable. The cost estimator, when he/she wants to use this model for estimating the cost of a new product, looks for the values of the inputs to be entered in this model.

It may happen, if the development of the project has been nearly completely done, that the product mass (to limit the discussion to this variable) is accurately forecasted – or it may even have been measured on a prototype – with an accuracy sufficient enough (let say one to a few percent) to properly use the model. The situation is simple.

However, an estimate is often required in the early stages of a project, during what we called the "decision phase". At this time, the physical characteristics of the product are not known yet: the only things engineers know are the functional requirements (for instance the product should be able to handle so many watts). This is a strong incentive to develop cost models using functional characteristics, but, if the cost analyst did use the mass, the cost estimator needs this information.

He/she therefore goes interviewing the engineers. If he/she directly asks for the value of the mass, it is rather sure that the engineers will answer: "we do not know; it is too early". What they mean at this time is that they cannot promise or make a guess for a single value, and they are certainly right. But the cost analyst may be sure that the engineers have a good idea of an approximate value of this mass. So the cost analyst should preferably ask: Is the order of magnitude of the mass about 100 kg or less than that? Depending on the answer, he/she continues until the engineers agrees that the mass will be between 20 and 25 kg: they know that the mass will get one value and only one and will not be subject to changes once it will be determined, but they cannot be more precise at this time. Sometimes some information may be added: in the opinion of the engineers, the mass will be closer to 20 kg than to 25.

How can the cost estimator use this interval for cost estimating? The use of probability would be completely wrong, as the mass is not going to randomly fluctuate between 20 and 25 kg during the product development and production! It will have one value, the problem being that we do not know inside this interval where the mass is going to be.

Conceptually the situation is close to handling imprecision of measurement in any engineering activity. The engineer wants to use a set of relationships which requires inputs. For some variables these inputs must be the result of measurements made on the experimental device he/she works on; the measurements are made with some imprecision. The engineer does not conclude that the "true" value will randomly fluctuate inside the interval given by the imprecision, but only that it is imprecise. He/she wants to take into account this imprecision in the computations.

### 18.2.2 Introduction to the "Fuzzy Numbers"

The term of fuzzy number derives from the theory of fuzzy sets, which has been known in the industry for about 15 years now. The purpose of this section is not to make a mathematical analysis of this theory but to present its capabilities to the reader.

*What Is a "Fuzzy" Set?*

In the theory of the sets, which is now a well-known mathematical subject, an element belongs or does not belong to a particular set; this is perfectly defined. The border of the sets is clear.

In current life, the situation that this mathematical theory describes does sometimes exist: we can say that an apple belongs to the set of apples. Bust most often the belongingness to a particular set is not so obvious. To remain in the domain of fruits, due to hybridization for example, it may be possible to create a fruit which is a bit an apple, and a bit a pear. The set of "apples" has now no clear border with the set of pears, as a fruit may belong – more or less – to both (Figure 18.1): this set is now "fuzzy"; this word is well chosen as it means that the border of the sets cannot be clearly distinguished.

# Introduction to "Risk" Analysis

Pure apple

Pure pear

**Figure 18.1** The border between "apple" and "pear" may be fuzzy.

This concept has to be more precisely defined. We start from objects belonging to an ordinary large set (for instance the "fruits"). Suppose we can define a property inside this set (for instance "being an apple"). As this property can more or less belong to any object $O$, we define, for any object, a number which quantifies the degree with which this object has the property. This number, which is defined in the interval[3] [0, 100], is represented by the symbol $\pi(O)$; for instance we may say that for a particular fruit the property "being an apple" is such that $\pi(O) = 70(\%)$. In the standard theory of set $\pi$ can only take a value equal to 0 or 1.

The set of the objects which have a $\pi$ different from 0 is said to be a fuzzy subset $S$ of the larger set. It can be said that the objects of this subset belong, more or less, to this subset, their degree of belongingness being quantified by the number $\pi$.

It is possible to expand this definition to several properties: if we define $n$ properties on the elements of the large set, we can distinguish $n$ fuzzy sets $S_1, S_2, \ldots, S_n$, and attach $n$ numbers $\pi_1, \pi_2, \ldots, \pi_n$ to any element. The important point to notice here is that there is no reason for having the sum of these $n$ numbers equal to 1 for the same object $O$. This is the first difference with the theory of probability.

Some definitions:

- A fuzzy set $S$ is said to be *normalized* if there is at least one $O$ such as $\pi(O) = 100$; all the fuzzy sets we will use are normalized.
- A *cut-off level* $\alpha$ of the fuzzy set $S$ is the subset, called $S_\alpha$, of all the elements of $S$ for which:

$$\pi(O) \geq \alpha$$

The cut of level 1 is called the "core"[4] of the fuzzy set: it is the set of the elements of $S$ which belong "totally" to $S$.

For a given fuzzy set $S$, it is possible to use these $\pi$ values in order to define a new measure, called the "**possibility**". This measure is defined as follows:

- The possibility of a given element $O$ is given by $\pi(O)$. If the objects are the ordinary numbers $x$, the function $\pi(x)$ is generally called in the literature the "possibility

---

[3] We deliberately avoid the interval [0, 1] because it reminds too much the probability theory and we want to male clear that we are NOT dealing with probabilities.
[4] This subset is generally called in the literature the "modal value" of the fuzzy set. We deliberately use a different word, in order to avoid any confusion between this theory and the theory of probabilities.
[5] The word "repartition" is also used in the probability theory, with another meaning. But we do not need it in this book.

distribution". The word was not very well chosen because it still fosters the confusion between possibility and probability; you may prefer – as we do so in order to avoid any confusion – to use another word,[5] such as "possibility repartition".

- The possibility of the whole set $S$, $\Pi(S)$, is given by[6]

$$\Pi(S) = Sup[\pi(O), \quad \text{with } O \in S]$$

*This is the fundamental difference[7] between the theories of probability and possibility.*

Due to this definition, possibilities cannot be added as probabilities do. Probabilities are additive: the probability of the sum of two disjoint sets $A$ and $B$ is given by the sum of their probabilities:

$$P(A \cup B) = P(A) + P(B) \quad \text{when } A \cap B = \varnothing$$

whereas the possibility of the sum of two fuzzy sets $P$ and $Q$ is given by:

$$\Pi(P \cup Q) = \text{Max}[\Pi(F), \Pi(Q)]$$

We will see later on the consequence of this definition. For the time being just note that this formula has some logic: if one set at least has a possibility equal to 100 to exist, then the junction $P \cup Q$ also has a possibility equal to 100 to exist!

## The Fuzzy Numbers

We can now apply this theory to the value of any kind of variable (such as the variables themselves or the estimates); in this case, the set of values is defined as the set of the real numbers $\Re$.

Let $S$ be a fuzzy subset of $\Re$; to each value $x$ of $S$ (supposed to be normalized) is attached a number $\pi(x)$ in the interval $[0, 100]$. This function $\pi(x)$ can take any form (see Figure 18.2).

To simplify and because we consider that using any other repartition is rather artificial (except in very special circumstances), we will select a special possibility repartition (but other repartitions are possible) which covers most of the problem in cost estimating: the trapezoidal repartition (see Figure 18.3). Such a repartition includes, as particular cases, rectangular and triangular repartitions.

This trapezoidal repartition is completely defined by four numbers $\{a, b, c, d\}$; the interval $[b, c]$ is known as the "*core*" of the repartition, $L = b - a$ is called the "*left spread*" and $R = d - c$ the "*right spread*"; some authors prefer to define a fuzzy number from their spread and therefore by the four numbers $\{b, c, L, R\}$. All the distributions we will consider now will have this (normalized) form.

The fuzzy set $S$ which is so defined, with its possibility repartition (do not forget this repartition is not a probability distribution), is called a fuzzy number;[8] the result of one measurement of this number, which will be an ordinary number, can give any value between $a$ and $d$.

---

[6] A normalized fuzzy set has always a possibility of 100.
[7] The axioms of both theories are the same, except for the additivity. The theory of possibility is therefore more general than the theory of probability: the second one is just a particular case of the first one.
[8] Some authors prefer to use here the word "fuzzy interval", reserving the word "fuzzy number" to the triangular repartition. As we do not make any difference here between the trapezoidal and the triangular (which is considered as a particular case of the first one) repartitions, we use the same word for both.

# Introduction to "Risk" Analysis

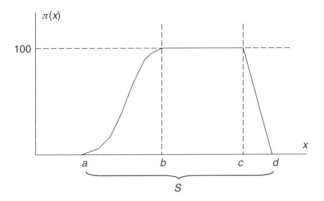

**Figure 18.2** Example of a normalized possibility repartition.

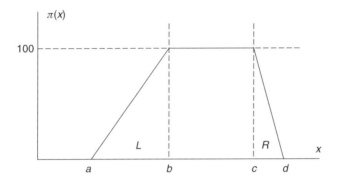

**Figure 18.3** The trapezoidal possibility repartition.

We must be very clear about this repartition: we are trying to estimate the value of something which does not exist yet, for instance the mass of an equipment. We know pretty well that this mass, once the development is finished, will have one value and just one. However, from the information we have today, the only thing we can say is that this mass will be somewhere between $a$ and $d$.

What is the meaning of spreads? They express opinions. In order to understand this sentence, let us return for a moment to the meaning of possibilities. Suppose we are in the open air, at a known distance of a building, and we have an optical instrument which allows us to measure angles, from which the size of the building can be determined. We make a first measure and compute 100 m. 100 m? Not exactly because our measurement has not an infinite accuracy: the only thing, we can say is that the building size is between $a$ and $d$, and that the "core" value is in the middle Figure 18.4. If we were in the fog, the spread may be larger.

In such a case, related to observations, the spreads just express the quality of the measurement. They are symmetrical. When guessing about the values a variable may have, the core is the interval for which no opinion can be given as all values are equally possible, whereas the spreads – which may be unsymmetrical – express opinions. This is similar to measurement imprecision, even if different words are used.

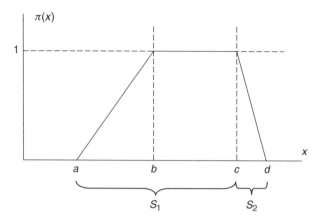

**Figure 18.5** Splitting $S$ into two subsets.

When we need an estimate for the mass of an equipment and interview engineers, they may say that they are sure that the mass will be between $a$ and $d$ but that, according to their opinion, the values between $c$ and $d$ are more possible than others: the possibility repartition represents this opinion.

### A Word of Caution

The consequence of the fundamental difference between the theories of probability and possibility is that, in the latter one, there is nothing like a "cumulative possibility": this has no meaning at all.

Suppose we split the set S into two subsets $S_1$ and $S_2$ (Figure 18.5).

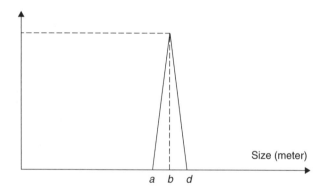

**Figure 18.4** Errors in measurement produces a "fuzzy" size.

It is impossible to say that the possibility of $S_1$ (to get $x$ in $S_1$) is higher than the possibility of $S_2$: as a matter of fact, both are equal to 100, because, according to the rule giving the possibility of a set, the maximum of possibilities inside the two subsets is equal to 100: therefore it is impossible to use the possibility repartition to decrease risk, by deciding to use a value as if, for instance, the mass is assumed to be $c$. As a matter of fact the mass can very well to be in the middle of the interval $[c, d]$: we do not know, even if we have an opinion about it.

## The Algebra of Fuzzy Numbers

The nice thing about fuzzy numbers is that it is possible to develop an algebra of fuzzy numbers: it is possible to add fuzzy numbers, to multiply them, to raise a fuzzy number to a fuzzy power, etc. If a cost estimating formula is built with this algebra, it will compute with fuzzy numbers at the same speed it computes with ordinary numbers (of course the result will be a fuzzy number as well): this explains why computing with fuzzy numbers is very quick and very efficient to deal with imprecision.

The algebra of fuzzy number is closely related to the propagation of errors in the computations.

This algebra can be rather complex. However, when the spreads are limited (let us say less than 20% of the value corresponding to the middle of the core), which is the case for dealing with variables imprecision, simple relationships were established by Jean-Marie Chauveau (Ref. [17]). These relationships are useful in cost estimating.

### Adding Two Fuzzy Numbers

This is a very frequent situation when one wants to use an additive formula, or to compute the total project cost when the costs of its elements are known.

Suppose we have two fuzzy numbers defined by:

$$X_1 = \{b_1, c_1, L_1, R_1\}$$
$$X_2 = \{b_2, c_2, L_2, R_2\}$$

Their sum is a fuzzy number defined as:

$$X_1 + X_2 = [b_1 + b_2, c_1 + c_2, L_1 + L_2, R_1 + R_2]$$

which is simple and logic.

### Subtracting Two Fuzzy Numbers

This operation is rather rare in cost estimating. It is simply given here to remind the reader that intuition should be handled with care. Nevertheless the result is also logic:

$$X_1 - X_2 = [b_1 - b_2, c_1 - c_2, L_1 + R_2, R_1 + L_2]$$

### Product of Two Fuzzy Numbers

The result is given by:

$$X_1 \times X_2 = [b_1 b_2, c_1 c_2, L_1 b_2 + L_2 \times b_1 - L_1 L_2, R_1 c_2 + R_2 c_1 + R_1 R_2]$$

which is less obvious!

A standard application in cost estimating is either the use of a multiplicative formula including several variables or the use of an additive formula where the variable is always multiplied by a constant. Applying the formula to the product of a fuzzy number by a constant $K$ simply gives the simple result:

$$K \times X = \{Kb, Kc, KL, KR\}$$

which is fortunately enough very logic.

*Raising a Fuzzy Number to a Power K*
This is a normal situation when using a multiplicative formula:

$$X^k = \{b^k,\ c^k,\ b^k - (b-L)^k,\ (c+R)^k - c^k\}$$

*Exponentiation of a Fuzzy Number*
The following formula is used when a exponential formula is preferred:

$$e^X = \{e^b,\ e^c,\ e^b - e^{b-L},\ e^{c+R} - e^c\}$$

This formula is logic.

It is also possible to compute the logarithm of a fuzzy number or the value of a fuzzy number at the power of a fuzzy number. These operations are not really used in cost estimating and the reader may consult Ref. [17] if such a problem arises.

### 18.2.3  Using Fuzzy Numbers Inside Cost Models

Models, specific or general, are set of formulae using addition, multiplication, exponentiation and logarithms. So the theory of possibility, for which any number is a set of four values, can easily be implemented in both models: they have just to be developed using the algebra presented in the previous section.

## 18.3  Using Discrete Probabilities

Discrete probabilities are (should be) the standard tool for dealing with uncertain, discrete, events which may appear in the future, such as:

- the results of the trade-off analysis – if it has not been performed yet – may imply a change;
- such or such material – to be decided after structural analysis – may have to be used;
- if the prototype does not give the expected results, such modification will have to be implemented;
- such or such contractor will be selected after studying his bid;
- and so on.

All these events are discrete by nature and are always part of a project involving some development and also often in other projects.

The decision tree is the normal way to take these uncertainties into account. The process is very simple: to each situation involving different issues, a probability is attached to each issue. Quite obviously such probabilities are basically subjective; fortunately (this fact is quoted in Ref. [40], p. 164) it has been shown by Lindley and Savage that degree of belief, which is really what is meant by subjective probabilities, obeys the axioms of probability and can therefore be dealt with the same way.

The principles of the method is illustrated in Figure 18.6. To each node are attached the cost of the next alternative and its probability. Once the tree is completely built, the cost and probability of each project branch (A, B, etc.) can be computed; the probability distribution of the total project cost can then be established.

# Introduction to "Risk" Analysis

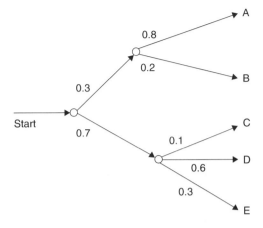

**Figure 18.6** Illustration of a decision tree.

## 18.4 Using the Continuous Probabilities

Using (continuous, as opposite to discrete) probabilities should be rather rare in cost estimating. The cost estimator must not forget that using probabilities distribution assumes that he/she believes that the value he/she is interested in will randomly fluctuate during the project life. If a cost driver does randomly fluctuate, the cost will also randomly fluctuate and the cost estimator will present to the decision-maker a probability distribution of the cost, based on the probability distribution of the value of the cost driver.

This is rather rare ... and we believe that most cost estimators use continuous probabilities because they believe it is the only tool available, but also because probabilities offer beautiful mathematical tools which allow them to present nice reports, even if they are not realistic, meaning that they do not properly describe the situation.

But it may happen. Probabilities were already used to compute the confidence interval of an estimate; using probabilities is there justified because the formula is computed on just one sample: drawing other samples might have given different CERs and therefore different estimates. As the samples was supposed to be randomly selected (this is the basis of all computations) probabilities naturally come into the picture. Probabilities may also have to be used if the project implies to manufacture a lot of similar equipments or parts, each one being subject to a lot of slight changes of specifications, changes which are not decided yet: the cost of one of each could practically be described by a probability distribution.

The reader is assumed to know the basis of the probabilities. A lot of books are available on the subject.

### What Are the Inputs?

To several inputs of the set[9] $X_1, X_2, \ldots, X_J$ is attached a probability distribution. The most important theoretical question is: Are the variable independent? If they are

---
[9] We use capital letters to remind the reader that we are estimating the cost of an object belonging to the population, not the sample.

several procedures can be used. If they are not, the Monte-Carlo simulation is probably the only procedure practically available.

The distributions can potentially be anything. However, we believe that selecting a too complex distribution is pure cosmetic (even if the computations can be carried out) and that no cosmetic should be allowed to the process which is sufficiently complex to understand – not to carry out – as it is. Therefore we recommend to use simple distribution which anyone can easily understand; consequently the most useful distribution could be the trapezoidal one, as it includes the rectangular and the triangular ones.

The range is supposed to be small in relative value, as the distributions are not there to say we know nothing about a variable, but that variable is known with some uncertainty. Let us say that a spread of not more than $\pm 20\%$ around the mode should be the rule (it is already a large range of uncertainty!).

## How Can Use This Information?

Let us assume that the cost is related to the inputs via a formula $\hat{Y} = F(X_1, X_2, \ldots, X_J)$. The problem we have to solve here is to get the distribution of a compound variable (the cost) which can involve adding variables, use variables at a non-integer power, etc.

Ideally we would like to solve analytically the problem. However, it is generally impossible as our mathematical tools do not allow it. Consequently we have to use approximations; approximations are largely sufficient in our domain as it would be strange to compute combined distributions with a full accuracy of variables known with an amount of uncertainty which could reach 20!

Among the approximate solutions, the most commonly used are the moments and the Monte-Carlo simulation.

### 18.4.1 Using the Moments

*Limitations*

The method of moments has two limitations:

1. The function $\hat{Y} = F(X_1, X_2, \ldots, X_J)$ is the sum or the product of linear function of one random variable. However, with some limitations generally met in practice, formulae using exponents (what we call in Volume 2 "multiplicative formulae") can be used thanks to the Mellin transform.
2. As defined in Section 18.1 of this chapter, the range of each random variable is limited. This is due to the fact that the theory of moments uses approximations, largely sufficient in practice, of distributions.

Both limitations, considered together, are not serious in practice.

*If You Are Using Specific Models (CERs)*
Of course there is no problem for the additive models.

For multiplicative models using one variable only, or the "correction by constant" model, the Mellin transform solves the problem. For more than one variable, the function must be linearly developed (using the Taylor's series) in the vicinity of

# Introduction to "Risk" Analysis

the nominal values of the variables. This transforms a multiplicative model using several variables into an additive model (of course true only in the vicinity of the nominal values, if limitation 2 (above) is fulfilled).

For other models the Taylor's series also solves the problem.

Once the moments are computed for each item, the project cost is just the sum of the items costs and limitation 1 is fulfilled.

### If You Are Using a General Model

We assume that you do not know the formulae used in the model.

In the vicinity of the nominal values, the variation of the cost in the vicinity of the nominal cost can easily be expressed as a formula: just compute the cost for three values of the variables (nominal, and ±10%) and compute the linear formula (or the parabolic formula if you want more precision) which goes through these values. You are back to the previous section.

## About the Moments of the Distribution of One Variable

Mathematicians showed that any distribution can be represented by an infinite set of moments, as a periodic function can be represented by an infinite set of sinusoidal functions (Fourier's series). Moments are numerated from 1 to ∞, the number related to one moment being called its "order" and will be noted by a index $k$. Practically we will limit the developments to the four first moments.

There are three types of moments:

- Centered moments, called $U_k$.
- Multiplicative moments, called $M_k$.
- Additive moments, called $A_k$.

It is always possible to convert one set of moments to another set. The conversion formulae are given below.

### The Centered Moments

Given the distribution of the variable $X$ represented by the function $p(X)$, the centered moments are defined by:

$$U_k = \int (X - \bar{X})^k p(X) dX$$

The moments are called "centered", because they are computed "around the mean $\bar{X}$.

The four centered moments are therefore given by:

$U_1 = 0$    by definition
$U_2 = S^2$    (variance or square of standard deviation)
$U_3$    allows to compute the level of skewness $\Gamma_1^2 = U_3^2/S^6$ or $\Gamma_1 = U_3/S^3$
$U_4$    allows to compute the level of kurtosis $\Gamma_2 = U_4/S^4$

## From Multiplicative to Centered Moments

$$U_1 = 0$$
$$U_2 = M_2 - M_1^2$$
$$U_3 = M_3 - 3M_1 M_2 + 2M_1^3$$
$$U_4 = M_4 - 4M_1 M_3 + 6M_1^2 M_2 - 3M_1^4$$

## From Additive to Centered Moments

$$U_1 = 0$$
$$U_2 = A_2$$
$$U_3 = A_3$$
$$U_4 = A_4 + 3A_2^2$$

## The Multiplicative Moments

The multiplicative moments, or moments in respect to the origin, are defined by:

$$M_k = \int X^k p(X) dX = E(X^k)$$

### From Centered to Multiplicative Moments
This requires to know the mean $\bar{X}$:

$$M_1 = \bar{X}$$
$$M_2 = U_2 + \bar{X}^2$$
$$M_3 = U_3 + 3U_2 \bar{X} + \bar{X}^3$$
$$M_4 = U_4 + 4U_3 \bar{X} + 6U_2 \bar{X}^2 + \bar{X}^4$$

### From Additive to Multiplicative Moments

$$M_1 = A_1$$
$$M_2 = A_2 + A_1^2$$
$$M_3 = A_3 + 3A_1 A_2 + A_1^3$$
$$M_4 = A_4 + 3A_2^2 + 4A_1 A_3 + 6A_1^2 A_2 + A_1^4$$

## The Additive Moments

These moments are extremely useful for computing the distribution of a project cost when the distribution of project elements are known.

There is no formula allowing to directly compute the additive moments: these moments are computed from the other moments.

# Introduction to "Risk" Analysis

## *From Centered to Additive Moments*
This requires to know the mean $\bar{X}$:

$$A_1 = \bar{X}$$
$$A_2 = U_2 = S^2$$
$$A_3 = U_3$$
$$A_4 = U_4 - 3U_2^2$$

## *From Multiplicative to Additive Moments*

$$A_1 = M_1$$
$$A_2 = M_2 - M_1^2$$
$$A_3 = M_3 - 3M_1 M_2 + 2M_1^3$$
$$A_4 = M_4 - 4M_1 M_3 + 12M_1^2 M_2 - 3M_2^2 - 6M_1^4$$

## *Using the Moments*

The interest of the moments is to be able to go directly from the distribution – its moments – of variables, to the distribution of their sum or their product.

All the results are based on the independence of the variables.

## *The Logic*
The computation is based on five steps:

1. Deciding on the distribution of the variables. In Chapter 3 of Volume 2 several distributions are presented; the user is recommended to use one of them.
2. Computing their moments. The same chapter presents the moments of all the distributions.
3. For each item of which at least one variable is defined as probabilistic, computing the moment of the cost distribution for this item. This is described in this section.
4. Computing the moments of the project cost distribution. The cost of the project is defined as the sum of the items cost: the computation uses the procedure described below.
5. Going from the moments of the cost distribution to its distribution. This is described in Returning to a distribution of this chapter.

## *Adding Several Independent Variables*
One shows that the additive moments of the sum are equal to the sum of the additive moments of the same order of the different variables:

$$\text{if } Y = \sum_n X_n \text{ then } A_k(Y) = \sum_n A_k(X_n)$$

The name of these moments comes from this property.

Consequently the additive moments of the sum $Y = X + B$ of one variable $X$ and one constant $B$ are given by:

$$A_1(Y) = A_1(X) + B$$
$$A_2(Y) = A_2(X)$$
$$A_3(Y) = A_3(X)$$
$$A_4(Y) = A_4(X)$$

## Multiplying Several Independent Variables

One shows that the multiplicative moments of a product are equal to the product of the multiplicative moments of the variables.

$$\text{if } Y = \prod_n X_n \text{ then } M_k(Y) = \prod_n M_k(X_n)$$

Consequently the multiplicative moments of the product $Y = X \times B$ of one variable $X$ and one constant $B$ are given by:

$$M_1(Y) = BM_1(X)$$
$$M_2(Y) = B^2 M_2(X)$$
$$M_3(Y) = B^3 M_3(X)$$
$$M_4(Y) = B^4 M_4(X)$$

## The Mellin Transform

Up to now, using the moments allows only to compute the moments of a sum and/or a product of random variables. This is sufficient if additive formulae are used. However, in cost estimating, we often use multiplicative or "correction by constant" formulae. This requires the Mellin transform (*source*: Ref. [30], p. 225) which is now described; this transform can only be used with a positive variable $0 < X < +\infty$ which is quite reasonable when dealing with cost.

### Definition

The distribution of variable $X$ is still given by $p(X)$. The Mellin transform of order $s$ of this distribution is a function of $s$ given by:

$$M_s^*(X) = \int_0^\infty X^{s-1} p(X) dX = E(X^{s-1})$$

which is nothing else that the expected value of $X^{s-1}$ (the notation $M^*$ is used here in order to avoid confusion with the moments). It can be said it is a generalization of the multiplicative moments. It can be immediately noticed that:

$$M_2^*(X) = M_1(X) \text{ (multiplicative moment of order 1, equal to } \bar{X})$$

$$M_3^*(X) = M_2(X) \text{ (multiplicative moment of order 2, equal to } U_2 + \bar{X^2} = S^2 + \bar{X}^2)$$

and so on.

For instance, the variance of $X$ is given by: $S^2 = M_3^*(X) - [M_2^*(X)]^2$

## Application to a Multiplicative Formula

Let us assume that $Y$ is defined by the following formula:

$$Y = b_1 \times X^{b_2}$$

where $X$ is for instance the mass supposed to be a random variable of which distribution is given by $p(X)$. The Mellin transform of $Y$ gives:

$$\begin{aligned} M_s^*(Y) = E(Y^{s-1}) &= \int Y^{s-1} p(Y) dY \\ &= b_1^{s-1} \times E((X^{b_2})^{s-1}) = b_1^{s-1} \times E(X^{b_2 s - b_2}) \\ &= b_1^{s-1} \times \int X^{b_2 s - b_2} p(X) dX \end{aligned}$$

which can be written:

$$M_s^*(Y) = b_1^{s-1} \times M_{b_2 s - b_2 + 1}^*(X)$$

Of course the procedure requires that the integral $\int_0^\infty X^{b_2 s - b_2} p(X) dX$ can be computed! It is not always obvious.

Once $M_s^*(Y)$ are known, it is possible to compute the multiplicative moments of $Y$, and from them all the other moments as required.

Two cases can be studied:

1. The case of a simple distribution of $X$, whatever the value of the exponent $b_2$: this is one of the most interesting case in practice.
2. The case of an exponent $b_2$ integer, whatever the distribution.

## The Distribution of X Is Relatively Simple, Whatever the Value of the Exponent $b_2$

If $X$ is uniformly distributed (rectangular distribution) in the interval $[l, h]$, the distribution is defined as:

$$p(X) = \frac{1}{h-l} \quad \text{if } l \leq X \leq h \quad \text{and} \quad p(X) = 0 \quad \text{elsewhere}$$

and consequently,

$$M_s^*(Y) = \frac{b_1^{s-1}}{h-l} \int_l^h X^{b_2 s - b_2} dX = \frac{b_1^{s-1}}{(b_2 s - b_2 + 1)(h-l)} \times [h^{b_2 s - b_2 + 1} - l^{b_2 s - b_2 + 1}]$$

If $X$ has a triangular distribution $[l, m, h]$, its distribution is defined by:

$$p(X) = \begin{cases} \dfrac{2(X-l)}{(h-l)(m-l)} & \text{if } l \leq X \leq m \\ \dfrac{2(h-X)}{(h-l)(h-m)} & \text{if } m \leq X \leq h \end{cases}$$

Consequently,

$$M_s^*(Y) = \frac{2 \times b_1^{s-1}}{(b_2 s - b_2 + 1)(b_2 s - b_2 + 2)(h - l)}$$
$$\times \left[ \frac{h(h^{b_2 s - b_2 + 1} - m^{b_2 s - b_2 + 1})}{h - m} - \frac{l(m^{b_2 s - b_2 + 1} - l^{b_2 s - b_2 + 1})}{m - l} \right]$$

Dealing with trapezoidal distribution is also possible, although more complex.

### $b_2$ Is an integer whatever the distribution

When the exponent is an integer, one goes directly from the moments of $X$ to the moments of $Y$:

$$M_s^*(Y) = b^{s-1} \times M_{b_2 s - b_2 + 1}^*(X)$$

which gives the following moments of $Y$:

$$M_1(Y) = M_2^*(Y) = b_1 \times M_{b_2+1}^*(X) = b \times M_{b_2}(X)$$
$$M_2(Y) = M_3^*(Y) = b^2 \times M_{2b_2+1}^*(X) = b_1^2 \times M_{2b_2}(X)$$

For instance if $b_2 = 2$, then

$$M_k(Y) = M_{k+1}^*(Y) = b_1^k \times M_{kb_2+1}^*(X) = b_1^k \times M_{2k}(X)$$

The multiplicative moment of order $k$ of $Y$ is equal to the multiplicative moment of order $2k$ of $X$ multiplied by a constant.

### Generalization

If $X_1, X_2, X_3$ are independent random variables and if the cost is given by:

$$Y = b_0 \times X_1^{b_1} \times X_2^{b_2} \times X_3^{b_3}$$

then the Mellin transform of order $s$ of $Y$ is the product of the Mellin transforms of $X_1, X_2, X_3$

$$M_s^*(Y) = b_0^{s-1} \times M_{b_1 s - b_1 + 1}^*(X_1) \times M_{b_2 s - b_2 + 1}^*(X_2) \times M_{b_3 s - b_3 + 1}^*(X_3)$$

### Computing the Moments

You may not be favorable to compute moments; it is true it is not a very interesting task! Fortunately some books have published moments for most used distributions; unfortunately these books published most often, as we do in Chapter 3 of Volume 2, moments of "normalized" distributions. Normalized distributions mean that the range of the variable is fixed to the interval [0, 1], whereas our costs are limited to the interval $[l, h] = [Y_{\min}, Y_{\max}]$.

# Introduction to "Risk" Analysis

Let us start from the distribution $p(\check{Y})$ of the normalized variable $\check{Y} \in [0, 1]$. This function becomes[10] $p(Y) = p(\check{Y})/(h-l)$ when $\check{Y}$ is replaced by $((Y-l)/(h-l))$ where $Y \in [h-l]$, as well as all values related to $Y$.

## The Rectangular (Uniform) Distribution

This distribution, under its normalized form is defined as:

$$p(\check{Y}) = 1 \quad \text{if } 0 \leq \check{Y} \leq 1$$

The denormalized form is written

$$p(Y) = \frac{1}{h-l} \quad \text{if } l \leq Y \leq h$$

## The Triangular Distribution

This distribution, under its normalized form, is defined as:

$$p(\check{Y}) = \begin{cases} \dfrac{2}{\check{m}}\check{Y} & \text{if } 0 \leq \check{Y} \leq \check{m} \\[2mm] 2\dfrac{\check{Y}-1}{\check{m}-1} & \text{if } \check{m} \leq \check{Y} \leq 1 \end{cases}$$

It becomes if $\check{Y}$ is replaced by $((Y-l)/(h-l))$ where $Y \in [h-l]$ and $\check{m}$ by $((m-l)/(h-l))$, taking into account the change of the distribution function:

$$p(Y) = \begin{cases} 2\dfrac{(Y-l)}{(h-l)(m-l)} & \text{if } l \leq Y \leq m \\[2mm] 2\dfrac{(h-Y)}{(h-l)(h-m)} & \text{if } m \leq Y \leq h \end{cases}$$

## Returning to a Distribution

Getting the moments of the cost distribution is fine, but it is extremely unlikely that the person who will receive the cost estimate will be able to interpret them. The cost analyst must present the distribution curve of the cost. The problem is then to go from the moments to the distribution.

Computation of the moments is limited to order 4. As we said before that any distribution can be represented by an infinite set of moments, it is clear that a large set of distributions may fit with only the first four moments. This means that we have to give another constraint. This constraint will be the shape of the distribution.

---

[10] This division has for effect to maintain the area under the distribution curve equal to 1.

Among all the shapes that could be chosen, our preference goes to the BETA distribution, because its range is limited and it has two independent parameters, called $\alpha$ and $\beta$, which are used to shape it.

## What Is the Function BETA?

The normalized (which means here that the area under the distribution is equal to 1) distribution BETA is a family of distributions defined on a finite interval $[l, h]$, where $l = Y_{min}$ and $h = Y_{max}$ are, for the time being, unknown:

$$\text{BETA}(Y; \alpha, \beta) = \frac{\Gamma(\alpha + \beta + 2)}{\Gamma(\alpha + 1)\Gamma(\beta + 1)} \left(\frac{Y - l}{h - l}\right)^\alpha \left(1 - \frac{Y - l}{h - l}\right)^\beta$$

where $\Gamma(u) = \int_0^\infty t^{u-1} e^{-t} dt$ is a generalization of the factorial function (if $u$ is an integer, $\Gamma(u + 1) = u!$) Its mode is given by $M = (\alpha h + \beta l)/(\alpha + \beta)$ and its mean by $\bar{Y} = l + (h - l)(\alpha + 1)/(\alpha + \beta + 2)$

The problem is to find out the four parameters $[l, h, \alpha, \beta]$ from the known moments of the $Y$ distribution. We first have to find the relationships between the moments and these four parameters. They are easily computed for the centered moments:

$$U_1 = 0$$

$$U_2 = (h - l)^2 \frac{(\alpha + 1)(\beta + 1)}{(\alpha + \beta + 3)(\alpha + \beta + 2)^2}$$

$$U_3 = (h - l)^3 \frac{-2(\alpha - \beta)(\alpha + 1)(\beta + 1)}{(\alpha + \beta + 3)(\alpha + \beta + 2)^3 (\alpha + \beta + 4)}$$

$$U_4 = (h - l)^4 \frac{3(\alpha + 1)(\beta + 1)[2(\alpha + \beta + 2)^2 + (\alpha + 1)(\beta + 1)(\alpha + \beta - 4)]}{(\alpha + \beta + 3)(\alpha + \beta + 2)^4 (\alpha + \beta + 4)(\alpha + \beta + 5)}$$

with

$$M = \frac{\alpha h + \beta l}{\alpha + \beta}$$

$$\bar{Y} = l + \frac{(h - l)(\alpha + 1)}{\alpha + \beta + 2}$$

From the centered moments, one can compute the following relationships based on the additive moments:

$$\frac{4(\alpha - \beta)^2 (\alpha + \beta + 3)}{(\alpha + 1)(\beta + 1)(\alpha + \beta + 4)^2} = \frac{A_3^2}{A_2^3}$$

# Introduction to "Risk" Analysis

$$\frac{3(\alpha+\beta+3)[2(\alpha+\beta+2)^2+(\alpha+1)(\beta+1)(\alpha+\beta-4)]}{(\alpha+1)(\beta+1)(\alpha+\beta+4)(\alpha+\beta+5)} = \frac{A_4}{A_2^2}+3$$

$$(h-l)^2 = A_2 \frac{(\alpha+\beta+2)^2(\alpha+\beta+3)}{(\alpha+1)(\beta+1)}$$

$$(A_1-l)^2 = A_2 \frac{(\alpha+1)(\alpha+\beta+3)}{\beta+1}$$

Once these four moments are known for the $Y$ distribution, we have to solve these four equations to get the value of the four parameters $[l, h, \alpha, \beta]$. The procedure uses the Newton-Raphson method which consists in computing, by iterations, the first two equations in order to get $\alpha$ and $\beta$:
Let us call:

$$B_3(\alpha, \beta) = \frac{4(\alpha-\beta)^2(\alpha+\beta+3)}{(\alpha+1)(\beta+1)(\alpha+\beta+4)^2}$$

$$B_4(\alpha, \beta) = \frac{3(\alpha+\beta+3)[2(\alpha+\beta+2)^2+(\alpha+1)(\beta+1)(\alpha+\beta-4)]}{(\alpha+1)(\beta+1)(\alpha+\beta+4)(\alpha+\beta+5)}$$

so that the equations can be written:

$$B_3(\alpha, \beta) = \frac{A_3^2}{A_2^3}$$

$$B_4(\alpha, \beta) = \frac{A_4}{A_2^2} + 3$$

Starting from estimates $\alpha_0, \beta_0$, one writes $\alpha_1 = \alpha_0 + \delta\alpha$ and $\beta_1 = \beta_0 + \delta\beta$ which allow to develop both equations in Taylor series:

$$B_3(\alpha_0, \beta_0) + \frac{\partial B_3(\alpha_0, \beta_0)}{\partial \alpha}\delta\alpha + \frac{\partial B_3(\alpha_0, \beta_0)}{\partial \beta}\delta\beta = \frac{A_3^2}{A_2^3}$$

$$B_4(\alpha_0, \beta_0) + \frac{\partial B_4(\alpha_0, \beta_0)}{\partial \alpha}\delta\alpha + \frac{\partial B_4(\alpha_0, \beta_0)}{\partial \beta}\delta\beta = \frac{A_4}{A_2^2} + 3$$

This system of two linear equations with two unknowns is easy to solve in $\delta\alpha$ and $\delta\beta$, from which a new couple $\alpha_1$ and $\beta_1$ allows to restart the procedure until it converges.
Once values $\alpha$ and $\beta$ are found, it is easy to get the values of $l$ and $h$.
The procedure is not too complex; the crucial point being, as it is always the case for this method, to get "good" initial values. The reader should however note that the procedure does not converge for any set of moments: this means that some functions of which moments are known cannot be adequately represented by a BETA function. In such a case the best choice is often to look for a normal distribution.

For finding initial values:

- if $A_3/A_2^2$ is small (let us say lower than 0.001), this means that $\alpha \approx \beta$ (the distribution is symmetrical) and the level of kurtosis is enough to compute their value:

$$\alpha = \beta = \frac{9 - 5 \times \left(\frac{A_4}{A_2^2} + 3\right)}{2 \times \frac{A_4}{A_2^2}}$$

- else, it is first necessary to determine which one of $\beta$ or $\alpha$ is the greater: if $A_3 > 0$, then $\beta > \alpha$, and the opposite if $A_3 < 0$. In this case, experience shows that:
  - if $(A^4/A_2^2) + 3 > 2.5$, then good initial values are 0.5 and 2,
  - otherwise $\dfrac{9 - 5 \times \left(\frac{A_4}{A_2^2} + 3\right)}{2 \times \frac{A_4}{A_2^2}}$ and $0.5 \dfrac{9 - 5 \times \left(\frac{A_4}{A_2^2} + 3\right)}{2 \times \frac{A_4}{A_2^2}}$ are good initial values.

From these initial values, the convergence is fast. For instance with $(A_3^2/A_2^3) = 0.663$, $(A^4/A_2^2) + 3 > 3.152$ and $A_3 > 0$, the sequence of iterations gives:

- $\alpha = 0.5$ and $\beta = 2$ (initial values)
- $\alpha = 0.215573$ and $\beta = 2.896392$ (first iteration)
- $\alpha = 0.295284$ and $\beta = 3.000707$ (second iteration)
- $\alpha = 0.30033$ and $\beta = 3.001727$ (third iteration) which finishes the process.

Once these values are computed, it is very easy to compute $l$ and $h$, which are the minimum and the maximum cost values and eventually to draw the distribution curve.

### 18.4.2 Using the Monte-Carlo Method

Monte-Carlo was a name given during the Second World War to a simulation method. The basic idea is simple: the cost is a function of several random variables which have each one a particular distribution. The function is too complex to be dealt with mathematically, even with the moments. Therefore instead of trying to solve it mathematically, let us simulate the behavior of the cost when the variables go over their ranges.

In order to do that, we randomly select a value for each variable: this gives us a set of values. Then we compute – analytically – the value of the cost corresponding to this set: this gives one cost value. Repeating this process a large number of times (thousands of them) will produce a large number of cost figures of which distribution can be displayed.

# Introduction to "Risk" Analysis

This is very simple and computers can easily be programmed for running the same program thousands of time, starting each time from a different set of values, randomly generated.

The purpose of this section is not to develop such a program, but just to indicate to the reader the process by which it is done. So the discussion is limited to the basic idea on which is based this process.

The only "problem" to solve is to generate these sets of randomly selected values of the causal variables, being sure that these variables follow predetermined probabilistic distributions. The solution needs two steps for each variable:

1. Generating a random number $r$ equally distributed (this means that its distribution is rectangular) between 0 and 1. This is not as trivial as one would expect but several programs do exist to do it properly and we will not discuss this point any further.
2. Taking into account the distribution curve of each variable, which is an input of the problem, for computing a value for this variable.

If this procedure is used one thousand of times, the distribution of the values should follow the given distribution curve.

If the distribution curve of the variable is rectangular between $l$ and $h$, the problem is easily solved: as $r$ range goes from 0 to 1, the value of the variable is simply given by: $X = l + (h - l) \times r$. Graphically this procedure takes the form given in Figure 18.7.

Starting from a random number between 0 and 1, as it is located on the vertical axis, this procedure generates another random number, equally distributed – which means here following a rectangular distribution – between $l$ and $h$.

The transfer function illustrated in Figure 18.7 is obviously the cumulative distribution of the variable. Consequently solving the problem involves:

2a - Computing the cumulative distribution of the variable $X$; this cumulative distribution gives, for each value of $X$ the probability $r$ to find a value between $X_{min}$ and $X$. Cumulative distributions of all the major distributions appear in the literature.
2b "Inverting" this cumulative distribution: knowing the value of $r$, how can the value of $X$ be found? This is equivalent to solving an equation and may require some computation.

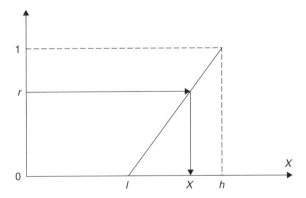

**Figure 18.7** From a random number between 0 and 1 to a random number between $l$ and $h$.

Let us take the example of a triangular distribution $[l, m, h]$. This distribution is defined by:

$$\frac{2(X-l)}{(h-l)(m-l)} \quad \text{if } l \leq X \leq m$$

$$\frac{2(h-X)}{(h-l)(h-m)} \quad \text{if } m \leq X \leq h$$

and the cumulative distribution is given by:

$$r = \begin{cases} 0 & \text{if } X < l \\ \dfrac{(X-l)^2}{(h-l)(m-l)} & \text{if } l \leq X \leq m \\ 1 - \dfrac{(h-X)^2}{(h-l)(h-m)} & \text{if } m \leq X \leq h \\ 1 & \text{if } X > h \end{cases}$$

which is illustrated in Figure 18.8 for $l = 10, m = 13$ and $h = 20$.

Now this distribution must be "inverted": knowing the value of $r$, what is the corresponding value of $X$? It is here not very difficult. For instance for $r = 0.2$, one easily compute $X = 12.45$. For this value of the variable, a cost can be computed. Repeating this operation thousands of time will give the distribution of the cost for the whole range of $X$.

This can be extended to any number of variables and even to dependent variables.

Some improvement can be made to improve the speed of the computations (for instance selecting random number more often in particular zones), but they have a limited interest.

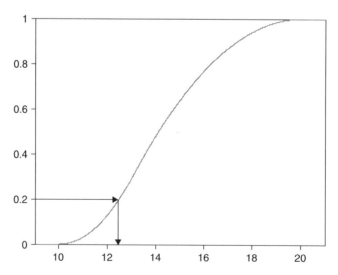

**Figure 18.8** The cumulative distribution of a triangular distribution.

Never forget when you use this procedure that you explicitly assume that the variables are random.

## 18.5 Mixing Both Methods

Can both approaches (using possibilities and probabilities) be mixed? Can we have a probability distribution of the cost which includes a possibility dispersion of the result?

The answer is yes, even it is very rarely used. It deserves some explanation because it helps explain the differences between both. As it is easier to explain using a Monte-Carlo simulation, this procedure will be used (but the moments can also be used: in such a case the possibility adjustments will be done once the distribution is computed).

The main problem to be solved when using both approaches is to present the results.

When using the Monte-Carlo simulation, the response of the program to one set of inputs (one value for the different causal variables) is deterministic; this may appear strange because we are dealing with probabilities but it is true: given a set of inputs, the value of the cost is clearly defined. Consequently the curve giving the probability distribution is just a sharp line, as illustrated in Figure 18.9.

Conversely to one probability corresponds one cost and only one.

This is not the case anymore when dealing with both possibility and probability approaches: now, as illustrated in Figure 18.10, to each probability correspond a fuzzy set of costs: to probability $p$ is associated a fuzzy cost number of which range is $C_1$ to $C_2$, and, as we see it in Section 18.2, it is impossible to propose any cost between these values. The cost is clearly fuzzy, whatever the probability.

Another way for saying the same thing is that the use of possibility when dealing with a probabilistic approach just "thickens" the line: no cost can be defined as a pure number.

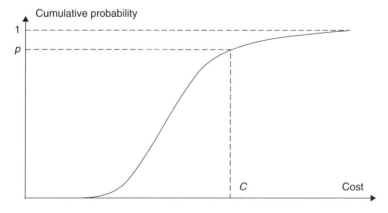

**Figure 18.9** The standard cumulative distribution of the cost.

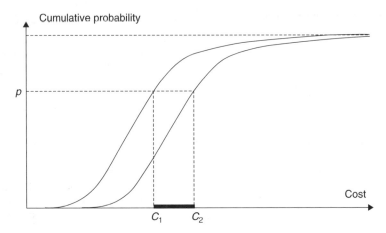

**Figure 18.10** To each probability is associated a cost defined as a fuzzy number.

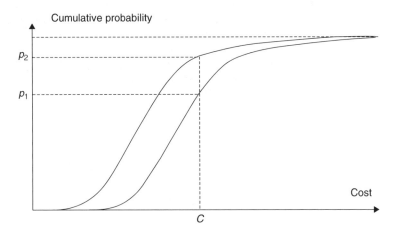

**Figure 18.11** To each cost is associated a probability range.

One could be tempted also to say that to one cost is attached a range of probability; this is explained in Figure 18.11: if the decision-maker asks you the probability of achieving cost $C$, you can only respond that the probability will be between $p_1$ and $p_2$.

# 19 Auditing Parametrics

## Summary

Auditing parametrics is rather trivial: it is like all other auditing process which are carried out in a company. It is after all normal that, from time to time, someone external to the cost estimating department has a look on the way cost estimates are produced. Cost estimates are prepared in order to make decisions; this means the company relies (at least partly) on them for its future: it is important!

One can distinguish three different types of audit:

1. *Looking at the models **utilization**.* This can occur frequently (let us say once a year), as it is the basis of the cost estimates. This is the job of the cost estimators.
2. *Looking at the models **preparation**.* This is the job of the cost analyst. Models are the tools which are going to be used. These tools are either made internally (generally speaking for the specific models) or bought from external sources (for general models). In the first case, one must check if these models are professionally made. In the second case, the cost analyst must prepare the usage of the model by establishing and communicating the rules to be observed inside the company on one hand, adjust the model to the company experience on the second hand. This should be done every 2 or 3 years.
3. *Looking at the **process**.* This is the job of the cost estimating department manager. All the steps of the process for preparing reliable cost estimates should be checked. This can be done every 5 years.

Auditing parametrics is an important subject inside organizations: if managers rely, in their decision-making process, on the figures given by the cost estimators, they must be confident that these figures correctly represent the (future) reality.

This confidence is built over the years. How is it built? There has been a lot of discussions about this subject among cost estimators, whatever the way the estimates are prepared.

Before the subject is discussed, it must be recalled that the main (if not the only one) reason to prepare an estimate is to help the decision-maker. This help can be done in two directions:

1. *Comparing alternate solutions.* Whatever the object of the decision, there are always different solutions to achieve a goal and the decision-maker has to choose

between them. One element, sometimes the most important one, which has to be taken into account is the relative cost of the solutions, the important word here being "relative": the cost estimator must be able to build a hierarchy between the solutions. It is less important here to give precise absolute cost figures than to give relative ones. In other words, if a bias does exist in the figures, it is not really important as long as this bias is the same for all of them (and of course as long as it is not too important, because the person who receives the cost estimates does not always make a clear distinction in his/her mind between absolute and relative figures). Nevertheless the main objective is to compare solutions. Among the examples, one can mention: trade-off analysis, design alternatives, materials selection, etc.

2. *Investment decisions.* Investment decisions imply allocating an amount of resources for a project. Now absolute figures are necessary. These figures are necessary for budgeting of course, but also for preparing the negotiation with a supplier; this is sometimes forgotten, but it must be reminded that the best way to prepare such a negotiation is to get a fair value of what the "object" – whatever it is – should cost. This means to be able to make an independent cost analysis ... If, during the negotiation phase, a change is made in the specifications, we are back to the previous paragraph; this illustrates the fact that it is always better to build an estimate on specifications rather than on technical description and that a cost model is an important tool during this phase, as time is not available to redo a whole detailed estimate.

A question which is frequently asked is "What is the accuracy of the cost/price estimate"? This question is natural: after all the person who receives the estimate will base his opinion about the work done by the estimator on the distance between the estimated cost and the "real" cost. Of course the measurement of this distance must be done on similar objects: this means – and the cost estimators are prone to mention this point – that there should not be too many technical changes in the object; otherwise a new estimate should be done with the new definition of the product (plus other factors such as the economic conditions, cost of raw materials, or other constraints).

Inside a company it is quite frequent to make a judgment on parametric cost estimating based on the comparison between parametric estimates and detailed estimates, the second ones being considered as "reference" not to be challenged. The origin of this fact is purely cultural: still nearly all managers consider that the "right" cost can only be done if the machines which are going to be used are defined, the operations sheets prepared, etc., in other words if the whole manufacturing/development process can be simulated. There is obviously some truth in this statement: if the manufacturing process can be correctly reproduced on the paper, the cost should be correct. But it completely forgets that the detailed cost estimators have, for any operation, to "guess" how much time will be required for such or such an operation: they have charts to help them, or maybe only their experience: in other words they are parametric cost estimators at a lower level! This process implies handling a lot of figures, a lot of "guesses", ... among which people can forget something or make mistakes. For this reason an overall estimate acts as a safe barrier: we are back to conventional parametric cost estimating. No method can be considered as the "reference": they sustain each other and any difference reveals a potential problem in one or the other method, potential problem which has to be explained.

If the comparison is relatively easy when dealing with costs, the controversy is much acute when dealing with prices. Cost estimators are prone to reject comparisons between their estimates and actual prices based on the true reason that they do not know the price policy of the supplier. Nevertheless going from cost to price is only partly a game: first it involves taking into account many things that the cost estimator did not take into account (such as the packaging and the transportation, sometimes the installation, the insurances, the cost of money, etc.) and then a part of a gamble.

Two comments, which always require to add some information, can be done in this respect:

1. If there is only one supplier, this supplier may have a history of working with your company; looking at this history may reveal a permanent bias between your own estimates and the prices. If there is no history, it must be recognized that the "gambling" is not random; it is based on three major components: the workload of the company, the amount of competition and the desire to enter (if this is the case) into a new market. Authors have already tried to simulate this process, with what seems to be a limited success. Nevertheless there is certainly a logic, which has to be at least partly guessed.
2. If there are several suppliers, the added information is there. First of all differences in the proposed technical solutions must be taken into account. Second it is always interesting to study the distribution of the prices; this distribution, if the number of bidders is a bit large, is generally well represented by a normal distribution. Tails can be eliminated due to the known fact that if you ask the same question to a group of people, there are always people who underestimate the question (unless someone found a new intelligent solution which is cheaper than current solution: this cannot be eliminated), and others who overestimate its difficulty (they may have reasons to do so and it could be interesting to discuss the point with them). In the middle of the distribution the components mentioned earlier certainly have an influence.

Whatever the discussion, cost estimators cannot refuse the comparisons between estimates and bids, even if they have to be made with caution. It is legitimate that managers expect figures not too far away from the bids, even if some adjustments (which can be done by the experienced cost estimators) have to be made.

## What Kind of Audits?

Auditing parametrics is not something which has to be done once every 5 or 10 years! It should be done on a regular basis in order to detect possible deviations. We distinguish here three different levels of audit, called: short term, medium term, long term:

1. *Short-term auditing* should be done once a year in order to be sure that the procedures are properly followed. A few estimates should be selected to make this audit, which should not take more than 1 day.
2. *Medium-term auditing* should be done regularly, let us say every 2 or 3 years.
3. *Long-term auditing* should be performed every 5 years.

Auditing parametrics is the only way managers – and other persons who receive and use cost estimates – can rely on the information they receive. The credibility, a very important term when looking at forecasts, of the cost estimates depends on it.

*Organizing the Audits*

Auditing should be organized and planned (otherwise it will never be done). The person who will be in charge, (this work can be delegated to an external company of which knowledge and experience in parametric cost estimating is recognized), must be designated.

The purpose of auditing is to make sure that the work is made according professional standard, not especially to check the value of the estimates.

## 19.1 Short-Term Auditing: About Using Models

Short-term auditing is devoted to the **application of the model** which was used for estimating the project(s).

The first question to be asked (which is not related to any model) is the most important one: What kind of information do you have about the product to be estimated? We must never forget that cost estimating in general and parametric cost estimating in particular are based on comparisons between products; comparison implies that some information is available. Parametric cost estimating has a major advantage over detailed estimates in the fact that the level of information can be more synthetic; but it must exist. Estimating from just a product name and a mass (as we saw it so often) is not the proper way, unless you are sure that the product is similar to an existing product you know; otherwise, the model will give you a cost but you do not know its validity.

The second important question is: Is the cost estimate properly documented? The purpose of this documentation is not to criticize the cost analyst, but to make him/her able to properly compare the "true" cost of the product (the cost which will be given to him/her as a return from experience) with the cost he/she estimated. It is very rare, especially if the cost estimator is involved in the estimates in the preliminary phases of a project, that the product which is manufactured is exactly the same, and is produced in the same conditions as the information the estimator used when preparing his/her estimate. For this reason he/she must prepare the basis of future comparisons (which will have to be carried out sometimes several years after the estimate is done).

A third question, which does not depend on the type of model which is used, is: How risk is taken into account? Is the difference between uncertainty about the inputs (which should be dealt with by the theory of possibilities) and the possible events which can occur during the project life cycle (which should be dealt with by the decision tree or – in rare circumstances – by the theory of probabilities) clearly understood?

### 19.1.1 Using Specific Models

A specific model is built for a product family; this has been repeated many times in this book (especially in Volume 2); each time we insisted on the required homogeneity of this product family. Therefore, the first question to be asked is: Can the product being estimated really be considered as a member of the same product family? The question is easy to answer for simple products (such as the spare parts), but might be difficult for products considered as a whole (for instance equipments or even sub-systems).

This question is difficult to answer at the equipment level if you do not have any information about the content and sometimes the technology used in the product. Using a formula even declared "for electronic products" is dangerous, because the electronic technology (which changes quite often) is an important cost driver (much more than the material for a mechanical product) and the cost does also depend on the functions fulfilled.

The second question to be asked is: Is this formula the right one for our type of product? Returning to the example of an electronic box, if the formula does not require any input regarding the technology, such a formula should be used with severe caution, unless you are sure that technology is the same for all the boxes from which the formula was built.

Another question to be asked, in order to make a cost estimate, specific values were entered in the formula: Are you sure the meanings you give to the variables are the same as the ones which were used by the cost analyst who built the formula?

Obviously the auditor should check the following points:

- Are the values of the variables inside the range of the values used for building the formula? It is generally risky, unless you get more information, to use a formula outside the range of the variables it was built with.
- When using several variables, are not some variables more or less collinear? In such a case you must investigate the way the cost analyst handled this problem and how it was solved.

How confident is the cost estimator about the values he/she used for describing the product? Should the concept of fuzzy number have been used instead of a precise value?

If any subjective variable was used: How was the value selected? Is it properly documented? What is the influence of the subjective variable on the cost?

### 19.1.2 Using General Models

When using a general model, the situation is different, as the model is supposed to be able to estimate the cost of anything, inside a class of industry.

The major problem when using a general model is what was called earlier the "translation problem". An engineer is perfectly able to "describe" a product, using his/her own words. The model builder used his/her own words – or the same words, maybe with a different meaning: To what extent does the cost estimator know the differences? The cost estimator interviews the engineers who respond with their own words. Back in his/her office he/she must "translate" these words in terms the model "understands" which means with the definition the model builder gave to them; any cost estimator using a general model knows it might be difficult.

The solution has two parts – besides a good understanding of the model vocabulary:

1. The cost estimator must be very consistent in his/her translation. A well-done model should therefore be able to present him/her with the meaning he/she gave to the words in past projects. This obviously assumes that was recorded!

2. The model was adjusted (most of general models have a variable dedicated to this purpose) taking into account the meaning given to the variables. This variable was quantified for a given product or for a given industrial culture: How are you sure that the value you are going to use correctly describes the project you are working on?

If the model presents several options, what was the option selected and for which reasons?

If the model uses subjective variables (a frequent one, when dealing with development cost, is the level of experience of the staff which is going to handle the project), the same question as for specific models must be asked and of course what is the sensitivity of these parameters on the cost.

## 19.2 Medium-Term Auditing: About Preparing Models

Medium-term auditing is dedicated to **the way the model was built** or was adjusted for cost estimating in the company.

We are here going to audit the work made by the person we called the "cost analyst". The cost analyst may be the same person as the cost estimator. Nevertheless both functions should be clearly distinguished.

### 19.2.1 About Specific Models

The work of the cost analyst is here to build the specific model from the database.

The first question to ask is: Did the cost analyst truly follow all the steps we recommend when a specific model is going to be built? Are the results properly documented? Let us remind the reader the steps which must be fully accomplished:

- About the database:
  - Are you sure all data points really belong to the same product family? Was the information collected about the products sufficient to respond to the question?
  - Is the product family homogeneous enough?
  - Are the costs properly normalized?
- Looking for outliers. Were some data points eliminated? How many (if more than 10% of the data points were eliminated, the homogeneity of the database may be seriously challenged)? Were some corrections made on the costs? Are the reasons properly documented?
- If several quantitative variables were used, are not some of them collinear? What was their level of correlation? How was this problem solved?
- How and why the variables were selected for building the formula?
- How and why was the type of formula selected? Same question for the metric. If the standard linear regression used, what was the value of the bias? Was it corrected and how?
- What are the results of the statistical tests?
- And so on.

Are the models given to the cost estimator fully documented? Was the product family properly described so that the model user can decide to use this model or another one, or has to use this model with caution because the product to be estimated is really at the "border" of the product family? Was the range of the variables properly defined?

Is the control of the model (feedback from actual cost) really done and documented?

## 19.2.2 Using General Models

General models are models which are bought from an external source. When a general model is used in the company, the cost analyst is the person who makes the model "ready to work" in the company: he/she makes the interface between the model seller (or builder) and the model user (the cost estimator).

This function should be recognized in the company as a truly independent and important function: all the companies are different, the general model is unique. Consequently there are always some adaptations to be made to the model:

- *About the vocabulary*: terms that the model uses must be properly defined and, if necessary, compared to the words used by the company.
- *Adjustment*: Was some adjustment(s) made to the model to palliate a permanent deviation?

Furthermore we recommended that the model was checked on past costs. Was this job properly done and documented? What was the average deviation between the observed costs and the estimated cost? and the standard deviation?

In any case, the cost analyst function is to prepare the manual which will be used as the reference manual in the company. In order to do it, the cost analyst is guided by the manual given by the model builder, but this manual has often to be interpreted (or even "translated") for the company estimates. Even if the cost analyst and the cost estimator are the same person, this work has to be done for traceability on one hand, for evolution on the other hand. Any person fulfilling both roles knows that the way he/she is working slowly changes with time. This is perfectly correct if these changes are documented.

## 19.3 Long-Term Auditing: About the Process

Long-term auditing is dedicated to a general survey of the cost estimating process, from the database to the feedback from finished projects, from tools selection and training to personnel careers, and of course to a general comparison between estimates and observed costs.

### Looking at Cost Estimating as a Process

Many people seem to believe that cost estimates depend on a specific tool only. Cost estimating is really a whole process and it can only be improved if it is looked upon as a process. Colonel Russel Logan (Air Force Pentagon Communications Agency) observes that *"cost estimating should be a corporate process – an essential one at that – and not something to be singled out for or subject to budgetary axing just to save costs. It is an essential part of your business or it is not. If not, any effort expended on estimating is meaningless"*.

A paper written by the US Software Engineering Institute (SEI) in May 1995 distinguishes six requisites for a reliable estimating process which must be applied in any domain:

1. a corporate memory (historic database);
2. a structured process for quantifying the characteristics of the object to be estimated;
3. mechanisms for extrapolating from demonstrated accomplishments on past projects;
4. audit trails (variables used for going from the object's characteristics to the cost are recorded and explained);
5. integrity in dealing with dictated costs and schedules (imposed answers are acceptable only when legitimate design-to-cost processes are followed);
6. data collection and feedback processes that foster capturing and correctly interpreting data from work performed.

Cost estimates are made by a person for another person to make a decision. This decision may have very large consequences for the business. Consequently the credibility of the cost estimate is something which should always be present in the cost estimator's mind.

A second important point which is linked to the previous one is that the person who receives the cost estimate must understand the relationship between the inputs (let us say at this stage the product description) and the cost; he/she wants to know how the cost is going to change if some input is changed. Once again parametrics is a very important help to reach this end.

## Model Selection

There are only a few models which are available on the market but we are not going to discuss here the merits of each, because these merits must be judged according to what is expected from them. The most important question before making a choice is: What is the type of information we generally have when an estimate is needed?

A model includes several variables, some models including a lot of them. The number of variables which are available is not a criteria in itself, and one can observe different opinions about it in the cost estimating community:

- Some cost estimators would like to have many variables at their disposal in order to answer questions which are sometimes asked. But if most of these variables are not known at the time most estimates are needed, their interest is limited.
- Others would prefer to get simple models and pretend they are confused when there are too many variables.

As usual both opinions are legitimate and depend mainly on the time the cost estimate is needed: in the early stages of a project (during the decision phase), as the information is limited, the cost estimators would prefer to get very few variables.

Later one (during the validation phase), more information is available and the cost estimator would like to use all of it.

It must not be forgotten that a general model is "general" and that no simple model can represent a complex reality. Maybe the solution could be to clearly indicate, when the first screen appears, what the user wants and organize the other screens for this wish: each screen would be simple, but the model will be complex.

# Bibliography

1. AFITEP. *Estimation des coûts d'un projet industriel.* AFNOR, 1995.
2. George Anderlohr. What production breaks cost? *Industrial Engineering,* September, 1969.
3. Claude Andrieux. Normes de temps. *Lavoisier,* 1983.
4. Robert N. Anthony. *Management Accounting. Text and Cases.* Richard D. Irwin, Inc., 1964.
5. P. W. Atkins. *Physical Chemistry.* Oxford University Press, 1990.
6. A. R. Ballman. *A tutorial application of the Anderlohr break-in production technique.* Westinghouse Electric Corporation, ILSD Engineering, Hunt Valley, MD.
7. Ralph M. Barnes. Motion and time study. *Design and Measurement of Work.* John Wiley & Sons, 1980.
8. Barry W. Boehm. *Software Engineering Economics.* Prentice-Hall, Inc., Englewood Cliffs, NJ, 1981.
9. S.A. Book. *The Aerospace Corporation's Capabilities in Cost and Risk Analysis.* June, 1994.
10. Carl de Boor. *A Practical Guide to Splines.* Springer, 2000.
11. Pierrine Bouchard (born Foussier). Parametrics for underground construction. *Proceedings of the International Society of Parametric Analysts, 22nd Annual Conference,* Noordwijk, 2000.
12. John Bowers. Assessing risk in major projects. *Proceedings of the International Society of Parametric Analysts, 10th Annual Conference,* 1988.
13. George Bozoki. An expert judgment based software sizing model. *Journal of Parametrics,* XIII(1), May, 1993.
14. David A. Belsley, Edwin Kuh and Roy E. Welsh. *Regression Diagnostics.* John Wiley & Sons, 1980.
15. Barry J. Brinker (ed.). *Emerging Practices in Cost Management.* Warren, Gorham & Lamont, 1990.
16. C. Burnet. *The effect of aircraft size and complexity on the production learning curve.* British Aerospace Public Limited Co., About 1975.
17. Jean-Marie Chauveau. *Valoriser l'imprécision des coûts en gestion de projet.* ESCP, 1991.
18. A. Chauvel, G. Fournier and C. Raimbault. *Manuel D'évaluation Économique Des Procédés.* Editions Technip, 2001.
19. *Defense System Management College Risk Assessment Techniques,* 1st edition. July, 1983.
20. Norman Draper and Harry Smith. *Applied Regression Analysis.* John Wiley & Sons, 1981.
21. D. Dubois. *Modèles mathématiques de l'imprécision et de l'incertitude en vue d'applications aux techniques d'aide à la décision.* Thèse d'état. Université de Grenoble, 1983.
22. D. Dubois and H. Prade. *Théorie des possibilités. Application à la représentation des connaissances en informatique.* Masson, 1988.
23. D. Dubois and H. Prade. Fuzzy real algebra: some results. *Fuzzy Sets and Systems,* 1979.
24. Bradley Efron and Robert J. Tibshirani. *An Introduction to the Bootstrap.* Chapman & Hall/CRC, 1993.
25. Merran Evans, Nicolas Hastings and Brian Peacock. *Statistical Distributions.* John Wiley & Sons, 2000.
26. R. E. Fairbairn and L. T. Twigg. Obtaining probabilities distributions of cost with approximations of general cost model functions. *Proceedings of the International Society of Parametric Analysts, 13th Annual Conference,* 1991.
27. Pierre Foussier and Jean-Marie Chauveau. Risk analysis. Are probabilities the right tool? *Proceedings of the International Society of Parametric Analysts, 16th Annual Conference,* Boston, 1994.
28. Larry D. Gahagan. A practical approach to conducting a cost risk analysis. *Proceedings of the International Society of Parametric Analysts, 13th Annual Conference,* 1991.
29. Paul F. Gallagher. *Parametric Estimating for Executives and Estimators.* Van Nostrand Reinhold Company, 1982.
30. Paul R. Garvey. *Probability Methods for Cost Estimating Analysis.* Marcel Dekker, Inc., 1999.
31. Gene H. Golub and Charles F. Van Loan. *Matrix Computations.* The John Hopkins University Press, 1996.
32. Glyn James. *Modern Engineering Mathematics,* 3rd edition. Prentice Hall, 2001.
33. Douglas T. Hicks. *Activity-Based Costing for Small and Mid-Sized Businesses. An Implementation Guide.* John Wiley & Sons, 1992.

34. J. Johnston. *Econometric Methods.* McGraw-Hill Book Company, 1960.
35. M. Lambert Joseph. Software estimation and prioritized hierarchies. *Proceedings of the International Society of Parametric Analysts,* V438–V445, 1986.
36. A. Lichnerowicz. *Algèbre et analyse linéaire.* Masson, 1955.
37. Donald W. Mackenzie. Price-based estimating risk model. *Proceedings of the International Society of Parametric Analysts, 13th Annual Conference,* 1991.
38. E. Malinvaud. *Statistical Methods of Econometrics.* North-Holland Publishing Company, 1980.
39. John Mandell. *The Statistical Analysis of Experimental Data.* Dover Publications, 1964.
40. Harry F. Martz and Ray A. Waller. *Bayesian Reliability Analysis.* John Wiley & Sons, 1982.
41. Matz, Curry and Frank. *Cost Accounting.* South-Western Publishing Company, 1967.
42. Gerald R. McNichols. An historical perspective on risk analysis. *Proceedings of the International Society of Parametric Analysts, 5th Annual Conference,* 1983.
43. Frederick Mosteller and John W. Tukey. Data analysis and regression. *A Second Course in Statistics.* Addison-Wesley Publishing Company, 1977.
44. National Association of Accountants. *Current application of direct costing.* Research Report, 1961.
45. Anthony J. Pettofrezzo. *Matrices and Transformations.* Dover Publications, 1966.
46. C. Radhakrishna Rao. *Linear Statistical Inference and Its Applications.* John Wiley & Sons, 2002.
47. David A. Ratkowsky. *Nonlinear Regression Modelling: A Unified Practical Approach.* UMI, 2002.
48. T. L. Saaty. *The Analytic Hierarchy Process.* McGraw-Hill, New York, NY, 1980.
49. Lothar Sachs. *Applied Statistics.* Springer-Verlag, 1984.
50. G. Saporta. *Probabilités, Analyse des données et Statistique.* Editions Technip, 1990.
51. Jun Shao and Dongsheng Tu. *The Jacknife and Bootstrap.* Springer, 1985.
52. Society of Manufacturing Engineers. *Tool and Manufacturing Engineers Handbook,* 3rd edition. McGraw-Hill Book Company, 1976.
53. P. Sprent and N.C. Smeeton. *Applied Nonparametric Statistical Methods.* Chapman & Hall/CRC, 2001.
54. Rodney D. Stewart. *Cost Estimating.* John Wiley & Sons, 1982.
55. Charles R. Symons. *Software Sizing and Estimating. Mk II FPA.*
56. Henri Theil. *Principles of Econometrics.* John Wiley & Sons, 1971.
57. US Department of Defense. *Armed Services Procurement Regulation Manual,* 1975 edition.
58. Michel Volle. *Analyse des données.* Economica, 1985.
59. Charles Winklehaus and Robert Michel. *Cost Engineering,* August, 1982.
60. Thomas H. Wonnacott and Ronald J. Wonnacot. *Statistique.* Economica, 1990.
61. L. Zadeh. Fuzzy sets as a basis for a theory of possibility. *Fuzzy Sets and Systems,* 1, 3–28, 1978.

# Index

ABC
  definition 26
  activity based costing 26
activity 21, 25, 26, 34, 36, 107
added value 21, 24, 51, 146, 162
analogy
  definition 34
  qualitative 35
  quantitative 35
analytic 34, 36
attribute 228
auditing 283, 285, 288, 289

BETA distribution 99, 276
Bootstrap 248
budget 9, 14, 18, 159, 254
building
  size 231
  quality level 189, 231
  localization 232

CAIV 8, 33
CEC 98
center
  dynamic 176, 243, 247
CER 35, 43, 53, 194, 255, 257
centralization 238
chance 19, 255, 258
characteristics
  functional 59, 63, 64, 68, 78, 197
  physical 63, 134, 197, 200
  technical 61, 202
Chilton 43, 141, 142, 250
coefficient 75
consistency 44, 61, 83, 91, 250
contribution 19
core 261
correlation 64, 67, 70, 72, 78, 169, 185, 236, 288

cost
  accounting 4, 24, 28, 34
  analyst 11, 14, 28, 85, 108
  centers 24
  definition 3, 89
  direct 16, 18, 19, 140
  estimating 41
  internal 14, 85
  measurement 4, 13, 15
  overhead 18, 21, 23, 146, 148
  specific 144, 193
  variable 14, 17, 140
cost improvement curve
  Crawford's curve 105, 117, 119
  De Jong 126
  Wright's law 105, 113, 119
costing
  absorption 20
  direct 19
  full 21
credibility 9, 45, 57, 198, 285
culture 176, 185, 218, 220

data
  base 84
  clarification 87
  collection 83
decentralization 239
DELPHI 92
depreciation 15, 147
descriptor 42, 63, 187

ECF 102
elastic limit 211
economic
  normalization 82, 95
  conditions 89, 95
energy 203, 204
entropy 204

estimate
  overview 31
  should be cost 7
  will be cost 7
estimating
  method 34
  elementary 47
ETF 199
exchange rate 95, 101
expenditures 15
expenses
  direct 18
  general and administrative 17
  indirect 18
expert
  judgment 90
  subjectivity 35
  system 41

forecasting 3, 7, 10
FPC 111
fuzzy number 260

heuristics 38

imprecision 243, 256
industrial culture 220
inflation 96
initial conditions 191
innovation 178
ITF 147, 203

Jackknife 248

learning rate 113

machinability 37, 180, 215, 216
materials 18, 31, 107, 158, 179, 201, 214, 216, 236
Mellin transform 272
modelization 41
model
  general 45
  specific 45
  type 1 189
  type 2 190
  type 3 194
moments
  additive 270
  centered 269

definition 269
multiplicative 270
Monte-Carlo 278
MTM 36

normalization 95, 131, 167

parameter 37, 42
parametric 11
plant
  load 149
  capacity 135
possibility 261
potential 25
price
  definition 6
probability 254
process 216
product
  advanced 203
  description 32, 179, 203
  size 199
  structure 203
project
  phases 7

quality 10, 53, 74, 246, 249
quality level 11, 32, 179, 188, 210, 213, 231

ratio 21, 40, 48
REC 97
residual 176, 227, 243
risk 253

sample 45, 75
software 225
star diagram 62

technical functions
  external 199
  internal 203
trade-off analysis 71, 202
trend 62, 108, 110, 228, 247
tunnel 232

uncertainty 256, 286

VE 179